Algorithms and Applications for Academic Search, Recommendation and Quantitative Association Rule Mining

RIVER PUBLISHERS SERIES IN AUTOMATION, CONTROL AND ROBOTICS

Series Editors

SRIKANTA PATNAIK
SOA University
Bhubaneswar, India

ISHWAR K. SETHI
Oakland University
USA

QUAN MIN ZHU
University of the West of England
UK

Advisor

TAREK SOBH
University of Bridgeport
USA

Indexing: All books published in this series are submitted to the Web of Science Book Citation Index (BkCI), to CrossRef and to Google Scholar.

The "River Publishers Series in Automation, Control and Robotics" is a series of comprehensive academic and professional books which focus on the theory and applications of automation, control and robotics. The series focuses on topics ranging from the theory and use of control systems, automation engineering, robotics and intelligent machines.

Books published in the series include research monographs, edited volumes, handbooks and textbooks. The books provide professionals, researchers, educators, and advanced students in the field with an invaluable insight into the latest research and developments.

Topics covered in the series include, but are by no means restricted to the following:

- Robots and Intelligent Machines
- Robotics
- Control Systems
- Control Theory
- Automation Engineering

For a list of other books in this series, visit www.riverpublishers.com

Algorithms and Applications for Academic Search, Recommendation and Quantitative Association Rule Mining

Emmanouil Amolochitis

University of Aalborg
Denmark

LONDON AND NEW YORK

Published 2018 by River Publishers
River Publishers
Alsbjergvej 10, 9260 Gistrup, Denmark
www.riverpublishers.com

Distributed exclusively by Routledge
4 Park Square, Milton Park, Abingdon, Oxon OX14 4RN
605 Third Avenue, New York, NY 10017, USA

First issued in paperback 2023

Algorithms and Applications for Academic Search, Recommendation and Quantitative Association Rule Mining / by Emmanouil Amolochitis.

Routledge is an imprint of the Taylor & Francis Group, an informa business

Publisher's Note
The publisher has gone to great lengths to ensure the quality of this reprint but points out that some imperfections in the original copies may be apparent.

While every effort is made to provide dependable information, the publisher, authors, and editors cannot be held responsible for any errors or omissions.

ISBN 13: 978-87-7022-984-5 (pbk)
ISBN 13: 978-87-93609-64-8 (hbk)
ISBN 13: 978-1-003-33714-0 (ebk)

Contents

Abstract

In the book, we present novel algorithms for academic search, recommendation, and association rule mining that have been developed and optimized for different commercial as well as academic purpose systems. Along with the design and implementation of algorithms, a major part of the work presented in the book involves the development of new systems both for commercial as well as for academic use. In the first part of the book, we introduce a novel hierarchical heuristic scheme for re-ranking academic publications retrieved from standard digital libraries. The scheme is based on the hierarchical combination of a custom implementation of the term frequency heuristic, a time-depreciated citation score, and a graph-theoretic computed score that relates the paper's index terms with each other. In order to evaluate the performance of the introduced algorithms, a meta-search engine has been designed and developed that submits user queries to standard digital repositories of academic publications and re-ranks the top-n results using the introduced hierarchical heuristic scheme. On the second part of the book, we describe the design of novel recommendation algorithms with application in different types of e-commerce systems. The newly introduced algorithms are a part of a developed Movie Recommendation system, the first such system to be commercially deployed in Greece by a major Triple Play services' provider. The initial version of the system uses a novel hybrid recommender (user, item, and content based) and provides daily recommendations to all active subscribers of the provider (currently more than 30,000). The recommenders that we are presenting are hybrid by nature, using an ensemble configuration of different content, user, as well as item-based recommenders in order to provide more accurate recommendation results. In the third part of the book, we present the design of a quantitative association rule mining algorithm. Quantitative association rules refer to a special type of association rules of the form: antecedent implies consequent consisting of a set of numerical or quantitative attributes. The introduced mining algorithm processes a specific number of user histories in order to generate a set of association rules with a minimally required support and confidence value. The generated rules show strong relationships that exist between the consequent and the antecedent of each rule, representing different items that have been consumed at specific price levels. This provides valuable knowledge that can be used for boosting the performance of recommender algorithms. We have introduced a post-processor that uses the generated association rules and improves the quality (in terms of recall) of the original recommendation functionality. The algorithm has been extensively tested on available production data, publicly available datasets, as well as custom-generated synthetic datasets simulating different market scenarios with the respective number of users and the respective number of transactions as well as fluctuation in prices depending on changes in demand.

Acknowledgements

My deepest gratitude goes to my supervisor, Prof. Ioannis T. Christou for his mentorship, during my Ph.D. studies and M.Sc. thesis, as well as for giving me the opportunity to collaborate on a series of state-of-the-art research projects. To my supervisor, Prof. Zheng-Hua Tan, go my deepest thanks for providing valuable insights in the field of Machine Learning.

This book is dedicated to my wife Sophia and my daughters; Melina and Lydia. You have given me hope, happiness and a new purpose in life $\sim$ I love you with all my heart.

List of Figures

List of Tables

1

1.1 Introduction

With the wide spread of the World Wide Web and the exponential growth of available specialized information, there is nowadays, more than ever, a need for efficient information retrieval solutions that aim to organize and process the vast amount of available data. In addition to the increase in data volume, available information has increased with respect to both complexity and semantic depth. General purpose search engines have made remarkable advances during the past years with respect to context identification, yet the need for specialized information retrieval systems seems hard to avoid.

For instance, online repositories like scientific libraries host an ever-increasing number of scientific publications, many of which tend to be interdisciplinary in nature, covering a wide range of topics, and at the same time, are hard to index using traditional classification schemes. The task of maintaining and updating the classification schemes, especially in ever-evolving fields such as that of science and technology, seems hard by itself. So, although indexing can be of considerable significance – and is a valuable component of most basic information retrieval methods – novel, sophisticated techniques are necessary to boost data retrieval efficiency.

Academic search engines have achieved some significant improvements during recent years. Still, at the time of writing, there is room of improvement and optimization, especially in cases of emerging scientific disciplines as well as specialized publications that deal with interdisciplinary topics of research. This proves to be a very challenging task, especially in cases of online libraries with a corpus of documents of considerable size as well as diversity. Furthermore, the increase in volume of available content makes the task of identifying newer, *trending* publications even more cumbersome, especially in areas where *authoritative* publications seem to prevail.

In addition to the aforementioned situation concerning the volume and nature of available content, there is also a significant increase in the number of available online web services that offer *consumable* content to users. The rate by which such online services are expanding and are being used by a continuously increasing number of subscribers makes recommender systems an emerging, promising area of research. Recommender systems aim to *push* personalized information – deemed of potential interest to specific users – based on prior knowledge, by means of historical data concerning the preferences of both the specific user – for whom recommendations are generated – and a wider group of potentially similar users.

Furthermore, the increase of available consumable content unavoidably results to a parallel increase in the number of available options for consumers, which makes price – in many cases – a determining factor with respect to the consuming behavior of a specific user. This introduces an interesting challenge in the design of recommendation systems, i.e., linking user preference with specific content considering user's sensitivity toward a certain maximum reference price that the user is most likely willing to pay. This information also provides a useful insight toward providing more attractive pricing schemes to potential consumers, eventually resulting in improvement in revenues for service providers as well a way for achieving customer retention.

During recent years, major companies in the search industry, including Google and Microsoft, have introduced some significant innovations in the field of academic search, launching search products, namely, Google Scholar and Microsoft Academic Search, that to a great extend have achieved efficient retrieval of academic publications online. Also a number of standard digital libraries, including Association for Computing Machinery (ACM) Portal and SpringerLink, have provided search solutions in an attempt to improve their online search functionality in order to facilitate the efficient retrieval of scientific publications located in their databases.

In addition to providing general purpose, as well as academic search functionality, companies like Google have expanded their efforts in areas beyond search. Specifically Google, having acquired YouTube, the online service for uploading and streaming user-submitted videos, offers recommendation functionality to subscribers of the service based on their watching history. Users of the service may also tag certain videos as favorable or not, which in turns affects the videos which are recommended to them. Having access to user ratings, as well as information concerning user behavior (percentage of total playing time watched among others) boosts the effectiveness of the recommendation process since there is a strong indication concerning user preference.

Although significant improvements and contributions have been introduced, there is still a big motivation for expanding knowledge and the current state-of-the-art in most of the aforementioned areas.

1.2 Algorithmic Motivation and Objectives

In the first part of the book, the focus is on the introduced methods for improvement on the search functionality provided by available academic search engines. Even though existing academic search engines have improved significantly during the last few years, being able to provide efficient results in response to complex queries remains an unsolved problem which attracts scientific interest. Being able to develop novel ranking algorithms for academic search engines would require a great effort with respect to obtaining a document database of such a size that would allow retrieval of relevant publications in response to an arbitrary number of diverse user queries from different fields. Therefore, in an attempt to limit the scope and focus of our research, our aim was to improve existing ranking algorithms, by introducing a meta-search engine system that aims to re-rank results retrieved from existing online search engines, in order to improve the quality of their top-n generated results. Also, although we limited our scope to publications dealing with the field of computer science and electrical engineering, the developed algorithms are applicable to any other scientific field, given that certain criteria are met as we will explain in a later section.

During recent years, considerable significance has been attributed to identifying *collaboration networks*, i.e., communities of scientists with common interests. By examining author co-authorship, and repeating the process for each author iteratively, one can form such networks of variable size and complexity. The idea of utilizing information concerning collaboration networks in different information retrieval areas has been attracting significant interest during past years. Our proposed methods examine the degree to which such networks of scientists can reveal common interests of different strength, and if so, identify the extent at which such information can be incorporated in the design of powerful ranking algorithms. This would require examining a document corpus of considerable size to be able to identify frequently co-occurring topics of interests as witnessed in the published work of scientists. Being able to identify such interests provides an insight about the overall relationships existing among different topics of research which might be part of interdisciplinary research work.

Another area that is extensively examined in the book is the identification of publications with the strongest affinity (content-wise) with the terms contained in a submitted search query. Using standard information retrieval heuristics such as term frequency-inverse document frequency (TF-IDF) was not possible, since the IDF part of the heuristic requires access to the entire corpus of publications, which is not commonly available to parties not affiliated with online repositories. This limitation introduced complexity in coming up with an efficient heuristic that can measure the degree of affinity among certain query terms and a specific publication.

Furthermore, another very important aspect was to be able to identify trending publications and promote those against older publications that may have higher citation score but are considerably older. This would allow "unearthing" publications that may be positioned lower in the rank based on absolute citation count values. The aforementioned approach gives credit to newer publications which might have lower citation count (in absolute value), but an emerging popularity that potentially makes the specific publication seem as more favorable than an older one with a higher citation count.

In the book, we also provide an overview of a commercial movie recommendation system that has been developed by the author for a major Greek triple play services provider. In the context of this book, it was required to design a system that incorporates novel recommender algorithms which are based solely on the users' watching history without having access to any input concerning user preference from a rating scheme or any other information such as the percentage of watching time for each item consumed by a specific user.

Apart from the absence of any additional information, save only the users' watching history, the recommender algorithms used by the system need to be able to address the issue that a single user account serves more than a single user. Specifically, most user accounts of the video-on-demand service are registered to a single household, which further connects a number of different viewers, belonging to different user categories which may have different preferences. So, the algorithms were required to be able to provide different recommendations based on different subsets of the user histories, which potentially correspond to different users bound with a single account.

The motivation was to develop such a system that addressed the aforementioned issues, and furthermore employed an ensemble of diverse recommenders (*item*, *content* and *user* based) which aimed to provide more accurate recommendation functionality compared to existing implementations of other similar algorithms. Furthermore, an additional constraint was that the system needs to have updated recommendations on a daily basis, requiring at the same time that the system is always responsive to recommendation requests and that it provides them using a minimum amount of the limited available resources.

As already mentioned, with the increasing number of available content items, it becomes apparent that a single user has a number of available choices as far as consumable content is concerned. This has as a logical implication that the users become very sensitive toward item pricing (considering the amount of available choices), which introduces an interesting aspect for recommender systems, that of being able to recommend items at a price that the user is most willing to pay. With respect to that, the motivation is to be able to examine user behavior by means of the relationships among different items consumed at specific price levels.

These relationships or association rules (of the form antecedent implies consequent) need to reflect the strongest (in terms of confidence) relationships between the antecedent and consequent of the rule by identifying the maximum price for which the consequent item may be consumed for a given minimum price at which the antecedent items are consumed. This information would give valuable insight with respect to not just the way different items are related but also the prices at which they are consumed.

1.3 Related Work

Graph-theoretic methods have been very popular in application in search algorithms. Since the early search engines, graph-theoretic methods have been developed and extensively used by general purpose search engines. For example, the influential Page-Rank algorithm used by the Google search engine is based on the link-structure of the web, and is deemed as one of the most powerful algorithms for identifying web pages considered to be *authorities* in their respective fields. This concept has been very influential for search algorithms and the core concept has been expanded to other areas of information retrieval, like, for example, in academic search.

So similar to the link structure of the web, additional methods based on social or academic collaboration networks have been used in citation analysis in Ma et al. (2008) in order to identify researchers which are considered to be "authorities" in Kirsch et al. (2006) in their respective fields. Additionally, Martinez-Bazan et al. (2007) have developed a graph database querying system that is aimed to perform information retrieval in social networks. Similarly, Newman (2001, 2004) have examined graphs depicting scientific collaboration networks with respect to structure to demonstrate collaboration patterns among different scientific fields, including the number of publications that authors write, their co-author network, as well as the distance between scientists in the network among others.

With respect to graph-theoretic models, the work performed by Harpale et al. (2010) is considered to be among the most relevant recent works in the literature. The authors have constructed CiteData, a collection of academic papers selected from CiteULike social tagging website's database and filtered through CiteSeer's database for cleaning meta-data regarding each paper. The specific dataset contains a rich link structure comprising of the references between papers as well as personalized queries and relevance feedback scores on the results of those queries obtained through various algorithms. The authors report that personalized search algorithms produce much better results than non-personalized algorithms for information retrieval in academic paper corpuses.

There have been additional attempts to model the strength of different relationships between collaborators of such networks. Specifically, Liben-Nowell (2007) use graph structures to examine the proximity of the members of social networks (represented as network vertices) which the authors claim that can help estimate the likelihood of new interactions occurring among network members in the future by examining the network topology alone. Furthermore, the community structure property of networks in which the vertices of the network form strong groups consisted of nodes with only looser connections has also been examined in order to identify such groups and the boundaries that define them, a concept based on the concept of centrality indices (Girvan et al., 2002).

In the same direction with an aim to examine the evolution as well as topology of collaboration networks, Barabsi et al. (2001) examined a number of journals from the fields of mathematics and neuroscience covering an 8-year period. The method consisted of empirical measurements that attempt to characterize the specific network at different points in time as well as a model for capturing the network's evolution in time in addition to numerical simulations. The combination of numerical and analytical results allowed the authors to identify the importance of internal links as far as scaling behavior and topology of the network are concerned.

Similar to the aforementioned approaches which aimed to examine collaboration networks of scientists with respect to structure, relationship strength, as well as topology, there have been attempts that aimed to examine the relationships among different topics of interest in the published works of scientists. Specifically, Aljaber et al. (2009) identify important topics covered by journal articles using citation information in combination with the original full-text in order to come up with relevant synonymous and related vocabulary to determine the context of a particular publication. This publication representation scheme, when used by the clustering algorithm that is presented in their paper, shows an improvement over both full-text as well as link-based clustering. Topic modeling integrated into the random walk framework for academic search has been shown to produce promising results and has been the basis of the academic search system ArnetMiner (http://arnetminer.org) (Tang et al., 2008). Relationships between documents in the context of their usage by specific users representing the relevance value of the document in a specific context rather than the document content can be identified by capturing data from user computer interface interactions (Campbell et al., 2007).

Many of the aforementioned approaches use information related to collaborating authors, as well as topics of interest, in order to be able to come up with sophisticated information retrieval algorithms that address a series of issues in academic search. There are different approaches in the current state-of-the-art; some methods utilize the graph structure and topology of the generated graphs, while others attempt to identify the presence of clusters in the graphs revealing patterns of collaboration. Furthermore, the application of such methods proves to be very powerful in order to identify patterns in the graphs which allow to perform more accurate predictions about the future with respect to collaborating authors or co-existing topics of interests in scientific publications.

Standard information retrieval techniques including TF are necessary but not sufficient technology for academic paper retrieval. Clustering algorithms prove to be also helpful in cases in order to determine the context of a particular publication by identifying relevant synonyms (or so-called *searchonyms*, see Attar and Fraenkel, 1977) and related vocabulary. It seems that the link structure of the academic papers literature as well as other (primal and derived) properties of the corpus should be used in order to enhance retrieval accuracy in an academic research search engine.

Similar to academic search engines, recommender systems have gained widespread popularity in recent years and are considered to have reached sufficient maturity as a technology (Jahrer et al., 2010; Ricci et al., 2011). The research performed in this particular field has started more than 20 years ago

(Goldberg et al., 1992; Shardanand and Maes, 1995), etc., and it focuses on examining different ways that recommendation systems can better identify user interests and preferences based on knowledge of the users' behavior as well as on characteristics of the items that they have consumed. Many different types of algorithms have been introduced (*content, item,* and *user* based), with each type focusing on different properties.

Contrary to the field of academic search (at least in *non-personalized search* context), a very common issue appearing in many commercial recommender systems is the fact that the systems are unable to promote in high positions results that happen to be of higher relevance to a specific user (based on the user's historical data), and in contrast, promote results which either happen to be trending well for the majority of users or are considered to be of higher popularity overall. Cha et al. (2007) spot such a behavior in the recommendation functionality of YouTube, as well as general purpose search engines. Whereas in general purpose search, such a behavior is anticipated, user-based and item-based collaborative filtering approaches should attempt to minimize this effect by using special formulae that promote less popular items when computing the user or item neighborhoods (see Karypis, 2001).

The performance evaluation of recommenders is deemed to be a very demanding task, since different approaches have been introduced. Shani and Gunawardana (2011) present a property-directed evaluation of recommendation systems attempting to explain how recommenders can be ranked with respect to properties such as diversity of recommendations, scalability, robustness, etc. In their work, they rank recommenders based on specific properties under the assumption that an improved handling of the property at focus will improve the overall user experience.

Also the datasets used in evaluating recommenders may have an impact on the performance of a recommender. Specifically, Herlocker et al. (2004) suggest that depending on the datasets used, different recommenders have displayed a variation in performance. Additionally, the authors note that a similar effect resulted using differently structured datasets. Dataset structure and size are also mentioned in Mild and Natter (2002) where the authors claim that dataset size (in terms of users) plays a significant role in the type of recommenders that should be used by a system. In their work, the authors also show that for a large dataset, linear regression with simple model selection provides improved results compared to collaborative filtering algorithms.

Similar to the use of information gained from scientific collaborative networks (which as we already saw, has gained momentum in academic search), collaborative filtering algorithms have been extensively used in various implementations of movie recommendation systems. Both user-based as well as item-based neighborhood exploration strategies met huge early success (for the first, the name "Collaborative Filtering" was coined early in the 90s) and have been applied in many different recommendation systems. Golbeck and Hendler (2006) present FilmTrust a system that combines information about the user's semantic web social network including information about networks peers, to generate movie recommendations. Similarly, Li et al. (2005) introduce a method that uses collaborative filtering approaches in e-commerce based on both users and items alike. They also show that collaborative filtering based on users is not successfully adaptive to datasets of users with different interests.

A very challenging issue in recommender systems research is for recommenders to address the issue of the absence of user ratings. The situation where user ratings are simply unavailable or non-existent makes the task of recommendation very challenging, since there are no direct indicator concerning user preference, and that kind of information should be implied by different types of information. For instance, Li et al. (2014), having no user ratings available in the dataset, present a novel one-class collaborative filtering recommender system that utilizes rich user information showing that such information can significantly enhance recommendation accuracy.

Collaborative filtering may prove to be very powerful, but many recommender systems are able to provide accurate recommendations by the use of content-based recommenders exclusively. For instance, Christou et al. (2012) present a system that uses a content-based recommendation approach in order to address the problem of finding interesting TV programs for users without requiring previous explicit profile setup, but by applying continuous profile adaptation via classifier ensembles trained on sliding time-windows to avoid topic drift. Similarly, Pazzani and Billsus (2007) focus on content-based recommenders and review different classification algorithms based on the idea that certain algorithms perform better when having specific data representation. The algorithms are used to build models for specific users based on both explicit information and relevance judgments submitted by users.

Association rule mining in the field of e-commerce is an idea that has been occasionally pursued during recent years and has been triggered by the success and popularity of *e-commerce* which has introduced massive databases of transactional data (Kotsiantis and Kanellopoulos, 2006). Association rule mining is considered as one of the most commonly used data mining techniques for e-commerce (Sarwar et al., 2000) and there have been different approaches introduced, all of which aim to optimize different aspects of the mining process in order to be able to provide more accurate recommendation results.

Lin et al. (2002) propose a mining algorithm for e-commerce systems that does not require prior specification of minimum required support value for the generation of the rules. On the contrary, they consider that by specifying a minimum required support, a rule mining system may end up with either too many or too few association rules which has a negative effect on the performance of a recommender system. The authors suggest an approach where they need to specify only a target range, in terms of number of association rules that such a system shall generate, and the system automatically determines the support value. The generated rules are mined for a specific user, reducing the mining processing time considerably, and associations between users as well as between items are employed in making recommendations.

Mobasher et al. (2001) describe a technique for performing scalable Web personalization after mining association rules from *clickstream* data from different sessions. In their introduced method, they use a custom data structure that is able to store frequent item sets and allows for efficient mining of association rules in real time without the need to generate all possible association rules from the frequent item sets. The authors state that their recommendation methodology improves effectiveness in terms of recommendation quality and has a computational advantage over certain approaches to collaborative filtering such as the *k*-nearest-neighbor.

Leung et al. (2006) introduce a collaborative filtering framework based on Fuzzy Association Rules and Multiple-level Similarity (FARAMS) which extends existing techniques by using fuzzy association rule mining taking advantage of product similarities in taxonomies to address data sparseness and non-transitive associations. The experimental results presented show that FARAMS improves prediction quality, compared to similar approaches.

Wong et al. (2001) introduce a novel approach for discovering and predicting web access patterns. Specifically, their introduced methodology (which takes into consideration various parameters, including the duration of a user session) is based on the case-based reasoning approach, and the main goal is to discover user access patterns by mining fuzzy association rules from the historical web log data. In order for the proposed method to perform fast matching of the rules, fuzzy index tree is used, and the system's performance is also enhanced using user profile data through an adaptation process. An effort for predicting user-browsing behavior using association-mining approach by Wang and Shao (2004) where they propose a new personalized recommendation method that integrates user clustering as well as association-mining techniques. In their work, the authors divide user session data into frames corresponding to specific time intervals, which are then clustered together in specific time-framed navigation sessions using a newly introduced method, called HBM (Hierarchical Bisecting Medoids) algorithm. The formed clusters are then analyzed using the association-mining method to establish a recommendation model for similar students in the future. They apply their introduced method to an e-learning web site and their results showed that the recommendation model built with user clustering by time-framed navigation sessions improves the recommendation services effectively.

Sarwar et al. (2000) examine methods and techniques for performing live product recommendations for customers, and they have developed several techniques for analyzing large scale data purchase data obtained from an e-commerce company, as well as user preference data from the MovieLens dataset. The recommendation generation process is divided into different sub-processes that include: representation of the input, formation of user neighborhoods, and finally the actual recommendation generation, which – among others – include association rules mining; specifically, they aim to discover associations between two sets of products such that the presence of some products in a particular transaction implies that products from the other set are also present in the same transaction.

1.4 Algorithmic Challenges

In the current book, a number of algorithmic contributions are presented which are applicable in different areas of data mining and information retrieval.

In the area of academic search, a heuristic hierarchical scheme is introduced which aims to improve the ranking quality of search engines for scientific publications developed for standard academic libraries such as ACM Portal, which contain certain classification schemes based on which publications can be efficiently indexed by authors. Specifically, our contribution aims to improve the ranking quality of a set of results generated by a default search engine, by actually re-ranking the top-*n* specified search results originally generated by the search engine. Our proposed ranking scheme is based on several different heuristic methods applied in a hierarchical configuration. Specifically, our scheme is based on a set of methods that are applied in a hierarchy that reflects the actual *strength* (or *significance*) of the heuristic algorithm at the specific level in being able to rank the results based on different publication criteria.

The proposed scheme contains three different heuristics applied in the following order: (i) TF, (ii) depreciated citation count (DCC), and (iii) maximal weighted cliques (MWC).

At the first level of the hierarchy, there is a custom implementation of the TF heuristic that, contrary to the default implementation of the heuristic which takes into consideration just the number of occurrences of the query terms in a publication, the introduced implementation considers different information such as term co-occurrences, as well as the distance of co-occurrences in different parts/levels of the publication (sentence, paragraph, section).

At the second level of the scheme hierarchy, lies a heuristic that aims to evaluate the DCC score for each publication. This particular score represents both the popularity of a particular publication with respect to the total number of citations received, but also aims to identify *trending* publications, i.e., publications with emerging popularity, and promote those against other publications that might have a higher citation count which has been achieved by virtue of popularity as well of an older publication date which allowed the accumulation of a higher citation count. The DCC score aims to *depreciate* citations received during older years, eventually emphasizing on the importance of publications received during latter years.

At the third level in the scheme hierarchy, there is a heuristic that evaluates the maximal weight clique matching score for a particular publication. During a preparatory stage, a scientific publication index term crawler has been developed that extracts index terms from a set of publications. A total of more than 10,000 publications have been extracted to then build a set of MWCs of weight above a certain threshold value. Then, for each publication in the set, the heuristic attempts to calculate the degree to which the index terms of a publication match to those of the established MWC and provide a score value that can be used for further ranking the results.

At each level in the hierarchy, a specific structure is provided as input containing an *ordered* set of search results generated (provided) by a *third-party search engine* in response to a specific query. The scheme is designed and implemented in a way, that at each level, a heuristic algorithm processes the aforementioned structure resulting in an updated version of the structure which contains all elements of the original set, but in a *possibly* different order, as determined by the heuristic method at the level. The output of each heuristic is then provided as input to the immediate next lower level in the hierarchy and is processed according to the aforementioned procedure.

Each heuristic algorithm in the hierarchical scheme processes the search results contained in the provided input structure, based on different properties of the scientific publication (relevant to the heuristic algorithm at the level), and places the results into buckets of different *range* sizes according to the score generated by the heuristic algorithm at each level. The number of buckets and the size of the bucket range have been determined empirically.

The ordering (ranking) of results based on different buckets aims to apply a strict policy which prohibits a heuristic that is lower in the hierarchy to *significantly* alter the ranking order of a set of search results that have been provided by a certain higher level heuristic. It is safe to say that a heuristic that is higher in the hierarchy *majorly* determines the final order of the results. The aforementioned principle is reflected in the *bucketing* logic, which aims to group together publications of similar *strength* with respect to a certain set of properties, relevant to specific heuristic. And in turn, each lower level heuristic follows, basically re-ranks the results contained within each bucket, and places them in even finer buckets that are passed to the immediate lower level for processing.

In the area of recommender systems, a fully parallelized ensemble of recommenders has been developed that allows for improved recommendation functionality. Specifically, an ensemble of hybrid, content, and user item predictor is used that is able to perform accurate recommendation predictions. We have design and development of AIT MOvie Recommendation (AMORE), a commercial movie recommendation system, the first such commercial movie recommendation system deployed in Greece by a major Triple Play services provider. AMORE has been developed as a REST web service in a service-oriented architecture. The AMORE web service expects recommendation requests by web service clients, based on pre-specified web service interfaces and upon web service call, calculates relevant recommendations that are returned to the client via web service responses.

In addition to the exposed web service which provides a set of methods, AMORE contains another component, the AMORE batch job which aims to facilitate the process of pre-caching recommendation results, that would allow the web service to retrieve cached recommendations with the minimum, most cost-effective number of operations.

In order to facilitate the caching process, the system uses two schemas, following the exact same data model (we would refer to those schemas as the *main* and *auxiliary* to distinguish among them). So as already mentioned, the purpose of the batch job is to maintain a constantly updated state of the recommendation data, reflecting the most updated estimated user recommendations based on the most recent user histories. In order to achieve this, the batch job aims to cache the recommendations generated for web service methods that are called most frequently, i.e., operations that are part of the core recommendation functionality, such as retrieving the top-n recommendations for a particular user. So the generated recommendations are cached and stored in persistence, and when a web service request arrives to the server, the server is able to retrieve and return the web service response by retrieving the already cached recommendations from persistence using the minimum number of operations (a simple *select* SQL operation). And furthermore, the caching operation is designed in a way that allows the system to be able to have *updated* results available which are refreshed within fixed configurable intervals. The rate at which new recommendations are generated and cached is determined by the system administrator. It makes sense to refresh the cached recommendations at intervals during which it is estimated that some minimal change in user transaction history may occur (which in sequence will cause an update in the list of generated recommendations). In the initial version of the system, recommendations are generated once per day. This has also been a business requirement, since the Movie Rental platform caches daily all user recommendations.

The batch process involves the following steps. First, the system aims to examine whether the back-end services of the provider are responsive. These back-end services provide information related to the subscribers of the movie rental service including their histories as well as the entire set of available items that are available for consumption. Once the system verifies the back-end modules' responsiveness, the system then calls the web service that retrieves the most recent, up-to-date transaction history for each of the active users of the service. The recent histories are then used as input to the recommender to generate updated recommendations for each active user of the service. Upon the completion of the aforementioned task, the batch process proceeds to the generation of top recommendations based on the transaction histories of all users.

Upon the completion of this final task, the system then calls a web service to notify the server that the process completed on the database schema referenced by the batch job so that the server will proceed with an update of its database reference to point to the schema containing the most recent recommendations.

By doing so, the server will always be able to return the recommendations that have been most recently added to the database, while at the same time, the batch process will proceed with the update of the auxiliary schema.

The system has been developed to be fully configurable with respect to the frequency by which the batch process runs, as well as additional parameters including the top-*n* recommendations to generate for each user among others. The web service uses a connection pooling mechanism that reads the database connection reference (which always corresponds to the last fully cached schema).

Additionally, the web service exposes a set of complementary methods for generating recommendations on-the-fly under different constraints. For example, one of the major issues that AMORE is facing is to be able to distinguish among different users that are possibly bound with a single account. This situation is very common, since many households which happen to be subscribers of the movie rental service have a number of different viewers bound to a single account. To address this situation, the web service has a set of methods which allow for specifying different parameters in order to be able to specify the time frame during which recommendations should be generated. By doing this, the system is able to generate recommendations corresponding to certain watching behaviors during specific hours of the day.

A very powerful aspect of recommendation systems is to be able to recommend items at prices that are deemed attractive to potential consumers. Specifically, the intention is to correlate user preference (in terms of content) with price and come up with relationships that link related items (as well as their purchase price) as evidenced in user transaction histories. These relationships, called quantitative association rules of the form *antecedent implies consequent* (where both antecedent and consequent are sets of item – price pairs), assume that if a certain user consumes all items contained in the rule's antecedent at a price level at least equal to the one specified in the antecedent for each item, then with a given *support* and *confidence* value, the rule can predict that the user will also consume the item that is part of the rule's consequent at a price level that is at least equal to the one specified in the consequent of the rule.

A post-processor has been implemented which aims to use association rules generated in order to improve the quality of the recommendations.

Specifically, the introduced post-processor uses a set of generated recommendations and applies a post-processing step by examining which of the generated association rules fire for each of the user, meaning the rules whose antecedent items have been consumed by a specific user at a price which is at least equal to the price specified. For those rules, the items contained in the rule's consequent are promoted only in case that the items have not been consumed by a user changing in this way the original recommendation list. The post-processor gives an extra weight to the recommendations that are part of the rule's consequent than the ones included in the recommendation list generated by the original recommender. In case that a recommendation contained in the post-processor is also part of the original recommendation list, then the number of positions that the specific recommendation is promoted up to the recommendation list is significantly higher, as an extra boost resulted by the increased confidence that the specific recommendation has been deemed relevant by both the recommender and some association rule.

The aforementioned method has revealed performance increase in terms of recall for the recommender containing the post-processor compared to the original recommender.

2

Academic Search Algorithms

2.1 Collecting Data from Scientific Publications

Associations among different topics of interest in the works of computer scientists have been examined, information that has been incorporated in the design of powerful ranking algorithms. In this direction, a web crawler has been implemented for retrieving basic information about scientific publications (such as the publication's authors, co-authors, year of publication, and index terms) in order to populate a database with the aforementioned data that could be later processed. Specifically, by crawling the ACM Portal web site, approximately 10,000 publications have been downloaded and all respective metadata have been extracted. The reason why ACM Portal has been chosen is that it contains a coherent scheme for authors to index their publications, which can be efficiently utilized. At the time of the writing, ACM used the 1998 version of the ACM Classification Scheme, which has been revised in 2012, but still, both schemes are for the time being supported by ACM Portal.

The crawler is initially provided with a number of influential, highly cited Computer Science authors which are considered to be *authorities* in their respective fields. For each of these authors, the crawler submits a search query via Google Scholar (which has the richest coverage in terms of scientific bibliography, and consequently, it has the best estimates of the paper's citation counts) to retrieve all publications published by the respective author. From the retrieved list, the crawler needs to process all those publications containing index terms (based on the ACM Classification Scheme), so all publication URLs not belonging to the ACM Portal are filtered out and are not processed. For all those publications belonging to ACM Portal, the application extracts and stores in persistence the publication's index terms, names of all authors, date of publication, citation count, as well as all ACM Portal publications

Figure 2.1   Flow of academic crawling process.

citing the current publication. All encountered authors that are not already processed by the crawler are stored in the database, to be processed at a following iteration. The flow of the process is visualized in Figure 2.1.

2.2 Topic Similarity Using Graphs

2.2.1 Graph Construction

After data have been collected from approximately 10,000 publications, two types of graphs have been constructed, each having a different type of semantic value.

2.2.2 Type I Graph

The strongest type of graph corresponds to the most direct relationship between index terms, namely, that of index terms coexisting in the same publication. So, in a Type I graph, two index terms t_1 and t_2 are connected by an edge (t_1, t_2) with weight w, if and only if there are exactly w papers in the collection indexed under both index terms t_1 and t_2.

Let $M: E \rightarrow R$ be a map containing as key an edge e and as value the edge's weight w_e for the specific type of association. Let P be the set of publications crawled for a specific period. Let G_1 be an *undirected* graph with initially *no* edges whose nodes are all the index terms covered in P.

1. Foreach publication p in P do

    a. Let T_p be the set of all index terms of p.
    b. foreach $t_p \in T_p$ do

        i. foreach $u_p \in T_p, u_p \neq t_p$ do

            1. if $e = (t_p, u_p) \notin G_1$ then

                a. add (t_p, u_p) in G_1.

          b. Set $M(e) = 1$.

       2. else Set $M(e) = M(e) + 1$.

       3. endif

    ii. endfor

  c. endfor

2. endfor

3. end.

2.2.3 Type II Graph

The next strongest type of graph involves index terms that happen to exist in different publications of the same author, but do not coexist in the same publication. Specifically, in a Type II graph, two index terms t_1 and t_2 are connected by an edge (t_1, t_2) with weight w, if and only if there are w *distinct* authors that have published at least one paper where t_1 appears but not t_2 and also at least one paper where t_2 appears but not t_1.

We construct the Type II graphs as follows. Let P be the set of publications crawled for a specific period. Let A be the set of all authors of publications in P. Let $G_2 = (V, E)$ be an *undirected* graph with initially *no* edges in E_2 whose node-sets V are all the index terms covered in P. Let $M: E \rightarrow R$ be a map containing as key an edge e and as value the edge's weight w_e for the specific type of association.

1. Foreach author a in A do

    a. Let $P_a = \{p | p \in A, p$ co-authored by $a\}$.

    b. Let $V_a = \{\}$.

    c. foreach $p \in P_a$ do

      i. foreach $u \in P_a, u \neq p$ do

        1. if $(p, u) \notin V_a$ then

          a. Let T_p be the set of index terms of p.

          b. Let T_u be the set of index terms of u.

          c. foreach $t \in T_p | t \notin T_u$ do

            i. foreach $r \in T_u | r \notin T_p$ do

              1. if $(r, t) \notin E_2$ then

                a. add $e = (r, t)$ in E.

                b. Set $M(e) = 1$.

               2. else Set $M(e)=M(e) + 1$.
               3. endif
           ii. endfor
       d. endfor
       e. add (p,u) in V_a.
      2. endif
    ii. endfor
   d. endfor
2. endfor
3. end.

2.3 Topic Similarity Using Graphs

Graphs of the aforementioned types have been constructed, covering different 5-year periods, in order to be able to model changes in associations of topics of interest in the time dimension. After the aforementioned graphs have been constructed, heavily connected clusters can be mined by computing all MWC in these graphs. The fact that the graphs are of limited size with only up to 300 nodes (each graph has only up to 13 node degree) addresses the issue of mining graphs being an intractable problem both in time and in space complexity. The problem is further reduced in complexity by considering edges whose weight exceeds a certain user-defined threshold w_0 (by default set to 5). Given these restrictions, the standard Bron–Kerbosch algorithm *with pivoting* (Bron and Kerbosch, 1973) applied to the restricted graph containing only those edges whose weight exceeds w_0 computes all maximally weighted cliques for all graphs in our databases in less than 1 min of CPU time on a standard commodity workstation.

2.4 System Architecture

The entire system architecture is depicted in the *Data Flow Diagram* in Figure 2.2. Overall, the system consists of seven different processes. Process P1 implements a focused crawler that crawls the ACM Portal in order to extract information about the relationships between authors who happen to have collaborated as well as the different topics they have worked on (as evidenced by the index terms used to *tag* their published work).

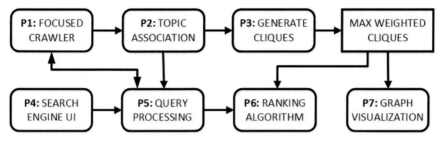

Figure 2.2 System architecture.

This information is analyzed in process P2 ("Analysis of topic associations and connections among authors and co-authors") and produces a set of edge-weighted graphs that connect index terms with each other. The process P3 ("Construction of maximum weighted cliques") computes fully connected subsets of nodes. The subsets form *cliques* that are an indirect measure of the *likelihood* that *a researcher working in an area described by a subset of the index terms in a clique might also be interested in the other index terms in the same clique.* All these cliques can be visualized via the components developed for the implementation of process P7 ("Interactive graph visualizations") using the Prefuse's Information Visualization Toolkit (Heer et al., 2005).

Processes P4-P6 form the heart of the prototype search engine we have developed, which includes a web-based application allowing the user (after registering to the site) to submit their queries. Each user query is then submitted to the ACM Portal and the prototype re-ranks the top-*n* ACM Portal results, and then returns the new top 10 results to the user. It is important to mention that in the testing and evaluation phase of the system, the results were returned to the user randomly re-ordered, along with a user feedback form via which the system got relevance feedback scores from the user, as explained in section.

2.5 Heuristic Hierarchy

The hierarchical scheme that has been introduced includes three heuristics, each located at a separate level in the overall hierarchy. The hierarchical structure of the configuration ensures that a heuristic at the top level in the hierarchy is considered as more significant in determining the final ranking of the results compared to a heuristic at a lower level. Therefore, heuristics are placed in a hierarchical structure to ensure that the ranking order is significantly determined by higher level heuristics but improved and fine-tuned by heuristics at lower levels.

There are three levels in our hierarchical heuristic scheme. At the first level, we have a custom implementation of the TF heuristic which aims to identify the degree at which specific query terms match the actual text context of a specific publication. Our implementation of the heuristic takes into consideration not just term occurrences, but details such as term co-occurrences in different levels (sentence, paragraph, and section) parts of the publications (title, abstract, and body). After calculating the TF score for each publication, based on the calculated value, the publication is placed in one of the pre-configured buckets, representing TF values of certain range size.

After the TF score is calculated and each of the available publications is placed in a bucket, the hierarchical scheme applies the second level heuristic: the DCC. DCC aims to estimate for each publication the degree of its *emerging* popularity. Specifically, the aim of the heuristic is to identify publications which have an increasing number of citations during recent years contrary to popular, older publications which have accumulated a significant number of citations over an extended course of several years. So, the heuristic basically depreciates the citation score based on the number of years lapsed since the paper has been cited. The heuristic applies on the buckets generated from the first heuristic and in sequence placed in finer grained second-level buckets.

At the third level in the hierarchy, lies the MWC heuristic which is applied on the two-level bucket structure filled by the second heuristic. Specifically, the MWC heuristic aims to find the matching degree between the index terms of each of the publications in the structure and each of the MWC stored in the database. The heuristic then sorts each of the publications in the two-level buckets based on the MWC score and ends up with a sorted list of results.

The heuristic hierarchy we use for *re-ranking the ACM Portal search results for a given query* is schematically shown in Figure 2.3.

2.5.1 Term Frequency Heuristic

As already mentioned, at the top-level of our hierarchical heuristic algorithm, a *custom implementation of the TF heuristic* is used. Term frequency is used as the primary heuristic in our scheme in order to identify the most relevant publications as far as pure content is concerned (for a detailed description of the now standard TF-IDF scheme; see, for example, Manning et al., 2009

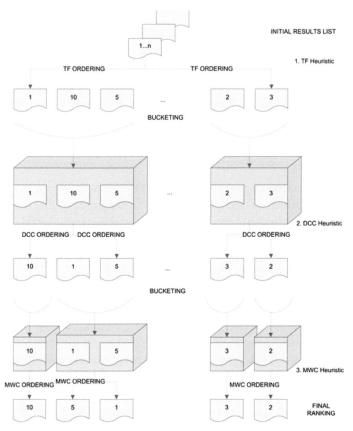

Figure 2.3 Re-ranking heuristic hierarchy.

or Jackson and Moulinier, 2002). When designing the TF heuristic, we have taken into consideration the fact that calculating the frequency of all terms individually does not provide an accurate measure for the relevance of a specific publication with respect to a specific query. To illustrate this, let's assume that for the query "distributed systems architecture," we have two publication results p_1 and p_2 with individual TF scores s_1 and s_2, respectively, where s_1 and s_2 are equal to the sum of the individual term frequencies for the query terms encountered in each publication. Let's also assume that $s_1 > s_2$, then based on the scores alone, the TF heuristic would assume that p_1 is more relevant than p_2 *ignoring* whether all or a subset of the query terms appear in each publication. So, in our example, p_1 might be strongly related to the topic "distributed systems" but have nothing to do with "distributed systems architecture" whereas p_2 might be a highly relevant "distributed systems architecture" publication, and yet p_1 would be considered more relevant publication.

In order to overcome this limitation, our implementation identifies the number of occurrences of all combinations of the query terms appearing in *close proximity* in different sections of each publication. After experimenting with different implementations of the TF heuristic, the experimental results showed that this approach performs significantly better in identifying relevant documents than the classical case of the sum of all individual term frequencies.

Our implementation assigns different weights to term occurrences appearing in different sections of the publication (Amolochitis et al., 2012) for results from an initial implementation that utilized the standard TF heuristic as described in most textbooks on Information Retrieval. Term occurrences in the title are more significant than term occurrences in the abstract, and similarly, term occurrences in the abstract are more significant than term occurrences in the publication body. Additionally, *the proximity level of the term occurrences is taken into consideration,* meaning the distance among encountered terms in different segments of the publication. By proximity level, we denote the distance among encountered terms in different segments of the publication, and for simplicity, we have two proximity levels: *sentence* and *paragraph*. Furthermore, the following two types of term occurrence completeness have been defined: *complete* and *partial*. A complete term occurrence is when all query terms appear together in the same proximity level, and similarly, a partial occurrence is when a strict subset of the query terms appears together in the same proximity level. The significance of a specific term occurrence is based on its completeness as well as the proximity level; complete term occurrences are more significant than partial ones and similarly term occurrences at sentence level are more significant than term occurrences at paragraph level.

Before discussing the details of our custom TF scheme, a word is in order to justify the omission of the "IDF" part from our scheme. The reason for omitting IDF is that we cannot maintain a full database of academic publications such as the ACM Digital Library (as we do not have any legal agreements with ACM) but instead fetch the results another engine provides (e.g., ACM Portal) and simply work with those results. It would be expected then that computing the IDF score for only the limited result-set that another engine returns would not improve the results of our proposed scheme and initial experiments with the TF scheme proved this intuition is correct.

We now return to the formal description of our custom TF scheme. Let $Q = \{T_1 \ldots T_n\}$ be the set of all terms in the original query, and let $O \subseteq Q$ be the subset of terms in Q appearing together in the same proximity level. We define the term occurrence score s_i for the ith term occurrence simply as $s_i = |O|/|Q|$. By ith occurrence, we denote the ith (co)occurrence of any of the original terms of Q in the publication. In case of a complete occurrence (meaning all query terms in the ith term occurrence appear in the original query as well) clearly, $s_i = 1$ since $O = Q$. Method calcTermOccurenceScore(O, Q) implements this formula.

Now, let T denote the set of all sections of a paper, P the set of all paragraphs in a section, and S the set of all sentences in a paragraph. The method splitSectionIntoParagraphs(*Section*) splits the specified section into a set of paragraphs. Similarly, splitParagraphIntoSentences(*Paragraph*) splits the specified paragraph into a set of sentences. The method findAllUniqueTermOccurInSentence(*Sentence*, Q) returns all unique occurrences of the query terms (that are members of Q) in the specified sentence. Similarly findAllUniqueTermOccurInAllSentences(S, Q) returns a set of all unique occurrences of the query terms (members of Q) in each sentence (members of S). The method noCompleteMatchExists(S) evaluates whether no complete term occurrence score exists in the sentences of S.

We have also introduced a set of weight values to apply a different significance to different term occurrence types appearing: (i) in different publication sections: *tWeight* represents the term occurrence weight at different publication sections (title, abstract, and body) and (ii) in different proximity levels: *sWeight* represents the term occurrence weight at sentence level, whereas *pWeight* represents the term occurrence weight at paragraph level. The method determineSectionWeight(t) determines the type of the specified section (title, abstract, or body) and returns a different weight score that should be applied in each case. All weight values have been determined empirically after experimenting with different weight value ranges. Overall, our term-frequency heuristic is implemented as follows:

Algorithm calculate TF (Publication d, Query q)

1. Let $S \leftarrow \{\}, T \leftarrow \{\}, P \leftarrow \{\}, O \leftarrow \{\}, tf \leftarrow 0$
2. Set $T \leftarrow$ splitPublicationIntoSections(d).
3. foreach section t in T do

    a. Let *sectionScore* $\leftarrow 0$.
    b. Set $P \leftarrow$ splitSectionIntoParagraphs(t).
    c. Let *scoreInSegment* $\leftarrow 0$.
    d. foreach paragraph p *in* P do

      i. Set $S \leftarrow$ splitParagraphIntoSentences(p).

     ii. Let *sentenceScore* $\leftarrow$ 0.

    iii. foreach sentence *s* in S do

        1. Set $O \leftarrow$ findAllUniqueTermOccurInSentence(s).

        2. Let *sScore* $\leftarrow$ calcTermOccurrenceScore(O, Q).

        3. Set *sentenceScore* $\leftarrow$ *sentenceScore* + *sScore*.

    iv. endfor

     v. Set *sentenceScore* $\leftarrow$ *sentenceScore* · *sWeight*.

    vi. Let *paragraphScore* $\leftarrow$ 0.

   vii. Let *partialMatch* $\leftarrow$ noCompleteMatchExists(S).

  viii. if (*partialMatch* === true) then

        1. Set $O \leftarrow$ findAllUniqueTermOccurrInAllSentences(S, Q).

        2. Set *paragraphScore* $\leftarrow$ calcTermOccurrenceScore(O, Q).

    ix. else Set *paragraphScore* $\leftarrow$ 1.

     x. endif.

    xi. Set *paragraphScore* $\leftarrow$ *paragraphScore* · *pWeight*.

   xii. Set *scoreInSegment* $\leftarrow$ *sentenceScore* + *paragraphScore*.

  e. endfor

  f. Let *tWeight* $\leftarrow$ determineSectionWeight(t).

  g. Set *sectionScore* $\leftarrow$ *tWeight* · *scoreInSegment*.

  h. Set *tf* $\leftarrow$ *tf* + *sectionScore*.

4. endfor

5. return *tf*.

6. end.

After calculating the total query TF for each publication, the algorithm groups all publications with similar TF scores into buckets of specified range. This grouping of the publications allows bringing together publications with similar TF scores in order to apply further heuristics to determine an improved ranking scheme. Results placed in higher range TF buckets are promoted at the expense of publications placed in lower TF buckets.

2.5.2 Depreciated Citation Count Heuristic

At the second level of our hierarchical ranking scheme, the results within each bucket created in the previous step are ordered according to a DCC score. Specifically, the annual citation distribution of a particular publication is analyzed, examining the number of citations that a paper has received within a specific year. We analyze all citations of a particular paper via Google Scholar, and for each citing publication, we consider the date of publication. After all citing publications are examined, we create a distribution of the total citation count that the cited publication received annually. Our formula then depreciates each annual citation count based on the years lapsed since the publication date. After all annual depreciation scores are calculated, the scores are summed and produce a total depreciation count score for a particular publication obeying the formulae:

$$c_p = \sum_{j=y(p)}^{n} n_{j,p} d_{j,p}$$

$$d_{j,p} = 1 - \frac{1 + \tanh\left(\frac{n-j-10}{4}\right)}{2}$$

(2.1)

where c_p is the total (time-depreciated) citation-based score for paper p, $n_{j,p}$ is the total number of citations that the paper has received in a particular year j, n is the current year, $d_{j,p}$ is the depreciation factor for the particular year j, and $y(p)$ is the publication year of the paper p. A graph of the citation depreciation function $d(x) = 1 - [1 + \tanh((x - 10)/4)]/2$ as a function of x is shown in Figure 2.4.

As already mentioned, the intention is to identify recent publications with high impact in their respective fields and promote them in the ranking order to the expense of older publications that might have a higher citation count but a considerable number of years have passed since the date of publication. To achieve this, the significance of a publication's citation count has been

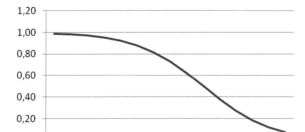

Figure 2.4   Annual depreciation of citation-count of a publication.

determined as a function of the number of citations received depreciated by the years lapsed since its publication date. Once publications have been sorted in decreasing order of the criterion c_p, the publications are further partitioned into second-level buckets of like-score publications.

2.5.3 Maximal Weighted Cliques Heuristic

Within each bucket of the second-level heuristic, the results are further ordered by examining each publication's index terms and calculate their degree of matching with all topical MWC, the off-line computation of which has already been described in Section 2.1.2. Additionally, specific weight values are assigned to the calculated cliques based on certain different characteristics such as the types of associations they represent and the time period they belong to. The system calculates for each publication a total clique matching score which corresponds to the sum of matching score of the publication's index terms with all MWC.

The calculation details are as follows.

Let C be the set of all cliques to examine. Let c_i denote the total number of index terms in clique i. Let d denote the total number of index terms of publication p and p_i denote the total number of index terms of publication p that belong to clique i; for each clique $i \in C$, the system calculates the matching degree of all publication index terms with those of a clique. In cases of a *perfect match* (meaning that all index terms of i appear as index terms of p) in order to avoid bias toward publications with a big number of index terms against cliques with a small number of index terms, we calculate the percentage match m_i as follows:

$$m_i = \frac{c_i}{d}$$

For all remaining cases (non-perfect match) the percentage matching is calculated using:

$$m_i = \frac{p_i}{c_i}$$

If $m_i > t$ where t is a configurable threshold for the accepted matching level (in our case $t = 0.75$) the process continues, else the system stops processing the current clique and moves to the next one. In case that the matching level is above t, the system calculates a weight score $w_{p,i}$ representing the overall value of the association of p with c_i as follows:

$$w_{p,i} = w_i \times m_i \times es \times ac_i$$

where w_i is the weight score of the examined MWC i and ac_i is a score related to the association type that the current graph that the current clique belongs to represents ($ac_i = 1$ for association type I and $ac_i = 0.6$ for type II). Finally, es is an exponential smoothing factor that depreciates cliques of graphs covering older periods in order to promote more recent ones. Since each type of graph has a different significance, we consider recent graphs of stronger association types as more significant and thus we assign greater value to MWC of such graphs.

The algorithm calculates for each publication a total clique matching score S_p which corresponds to the sum of matching score of the publication's index terms with all MWC and determines the final ranking of the results accordingly.

$$S_p = \sum_{i \in C} w_{p,i}$$

The total clique matching score determines the order of the results within the current second level bucket and eventually determines the final ranking of the results.

2.6 Experiments' Design

As previously mentioned, a meta-search engine application has been developed in order to evaluate our ranking algorithm. Registered users can submit a number of queries via our meta-search engine's user interface. The search interface allows users to use quotes for specifying exact sequence of terms in cases that it is applicable for improving query accuracy for both PubSearch and ACM Portal.

For each query in the processing queue, the system queries ACM Portal using the exact query phrase submitted by the user and crawls ACM Portal's result page in order to extract the top 10 search results. The top 10 search results as well as the default ranking order provided by ACM Portal are stored. For each of the returned results, the system automatically crawls each publication's summary page in order to extract all required information. Additionally, for each of the returned results, the system queries Google Scholar to extract the total number of citations and find a downloadable copy of the full publication text if possible.

When all available publication information is gathered, the system executes our own ranking algorithm with the goal of improving the default rank by re-ranking the default top 10 results provided by ACM Portal. The rank order generated by our algorithm is stored in the database, and when the process is complete, the query status is updated and the user is notified in order to provide feedback. The user is presented with the default top 10 results produced by ACM Portal in a random order and is asked to provide feedback based on the relevance of each search result with respect to the user's preference and overall information need. The provided relevance feedback score for each result is used for evaluating the overall feedback score of both ACM Portal and our own algorithm, since both systems attempt to process the same set of results. A 1-to-5 feedback score scheme is used, where 1 corresponds to "least relevant" and 5 corresponds "most relevant."

In order to compare an information retrieval system's ranking performance, two commonly encountered metrics are used: (i) Normalized Discounted Cumulative Gain (NDCG) and (ii) Expected Reciprocal Rank (ERR). A *new* metric, the *lexicographic ordering metric* (LEX), is introduced that can be considered a more *extreme* version of the ERR metric.

Normalized Discounted Cumulative Gain (Järvelin and Kekäläinen, 2000) is a metric commonly used for evaluating ranking algorithms in cases where graded relevance judgments exist. Discounted Cumulative Gain (DCG) measures the usefulness of a document based on its rank position. DCG is calculated as follows:

$$DCG_p = \sum_{i=1}^{p} \frac{2^{f(p_i)} - 1}{\log_2(1 + i)} \qquad (2.2)$$

where $f(p_i)$ is the relevance judgment (user relevance feedback) of the result at position i. The DCG score is then normalized by dividing it with its ideal score which is the DCG score for the sorted result list on descending based on the relevance scores resulting in:

$$nDCG_p = \frac{DCG_p}{IDCG_p} \qquad (2.3)$$

The term $IDCG_p$ (acronym for "Ideal DCG till position p") is the DCG_p value of the result list ordered in descending order of relevance feedback, so that in a perfect ranking algorithm, $nDCG_p$ will always equal 1.0 for all positions of the list. Expected Reciprocal Rank (Chapelle et al., 2009) is a metric that attempts to compute the *expectation of the inverse of the rank position in which the user locates the document they need* (so that when, for example, ERR = 0.2, the required document should be found near the 5th position in the list of search results), assuming that after the user locates the document they need, they stop looking further down the list of results. ERR is defined as follows:

$$\text{ERR}(q) = \sum_{r=1}^{n} \frac{R_r}{r} \prod_{i=1}^{r-1} (1 - R_i), \qquad R_i = \frac{2^{f(p_i)-1} - 1}{2^{f_{\max}}}, \qquad i = 1 \ldots n$$

(2.4)

where $f_{\max}$ is the maximum value the user relevance feedback score (in our case, 5).

Besides the common NDCG and ERR metrics, we also calculate a total feedback score LEX(q) for the (re-) ranked results of any particular query q by following a lexicographic ordering approach to produce a weighted sum of all independent feedback result scores:

$$\text{LEX}(q) = \frac{\sum_{i=1}^{n} a^i f_{\text{norm}}(p_i)}{\sum_{i=1}^{n} a^i}$$

(2.5)

where n is the number of results, $\delta_f = (f_{\max} - 1)^{-1}$, $a = \frac{\delta_f}{1+\delta_f}$, and $f_{\text{norm}}(p_i) = \frac{f(p_i)-1}{f_{\max}-1}$ is the normalized relevance feedback provided by the user for the publication p_i with values in the set $\{0, \delta_f, 2\delta_f, \ldots 1\}$. In our case, $\delta_f = 0.25, a = 0.2$. In this way, in any two rankings of some results' list produced by two different schemes, the scheme that assigns a higher score for the highest ranked publication always receives a better overall score LEX(q) regardless of how good or bad the publications in lower positions score. To see why this is so, ignoring the normalizing denominator constant in (5), and without loss of generality, we must simply show that if two result-lists $(r_{1,1}, \ldots r_{1,n})$ and $(r_{2,1}, \ldots r_{2,n})$ for the same query q get normalized feedback scores $(f_{\text{norm}}(r_{i,1}), \ldots f_{\text{norm}}(r_{i,n})), i = 1, 2$ and $f_{\text{norm}}(r_{1,1}) > f_{\text{norm}}(r_{2,1})$, then the LEX score of the first result list will always be greater than the LEX score of the second result list. Given that if two normalized feedback scores are different, their absolute difference will be at least equal to δ_f, and at most equal to 1, we need to show that

$$\delta_f a > \sum_{i=2}^{n} a^i \left[f_{\text{norm}}(r_{2,i}) - f_{\text{norm}}(r_{1,i}) \right]$$

(2.6)

for all possible values of the quantities $f_{\text{norm}}(r_{1,i}), f_{\text{norm}}(r_{2,i}), i = 2, \ldots n$. Taking into account that $f_{\text{norm}}(r_{2,i}) - f_{\text{norm}}(r_{1,i}) \leq 1, \forall i = 2 \ldots n$, if the value a is such so that $\delta_f a > \sum_{i=2}^{n} a^i = \frac{a^{n+1}-a^2}{a-1}$, then the required inequality (2.6) will hold for all possible values of the quantities $f_{norm}(r_{1,i}), f_{norm}(r_{2,i}), i = 2, \ldots n$. But the last inequality can be written as $\delta_f > \frac{a(1-a^{n-1})}{1-a}$ and it will always hold if $\delta_f \geq \frac{a}{1-a}$ (since $a^{n-1} \in (0,1)$), so by choosing $\delta_f = \frac{a}{1-a} \Leftrightarrow a = \frac{\delta_f}{1+\delta_f} = 0.2$, the lexicographic ordering property always holds regardless of the result list size or feedback values. Clearly, it always holds that $\mathrm{LEX}(q) \in [0,1]$, with the value 1 being assigned to a result list where all papers were assigned the value f_{max}, whereas if the user assigns the lowest possible score (2.1) for all papers in the results' list, the LEX score for the query will be zero. Also, notice that if the user assigns the median value $\frac{f_{max}+1}{2} = 3$ to all papers in the results' list for a query, the LEX score for that query will also be the median value 0.5.

The LEX scoring scheme can be considered as a more *extreme* version of the ERR and NDGC metrics and is inspired from the fact that people always place much more importance to the top results (and usually judge the whole list of results by the quality of the top 2–3 results) that are returned from any search engine than on lower ranked results. This is probably due to the very strong faith of users in the ability of search engines to rank results correctly and place the most relevant results on top, a faith that (if it exists) apparently does not have solid grounding with regard to *academic* search engines – at least, not yet.

2.7 Experimental Results

In an initial training phase, the results of a limited set of relevance feedback scores from a limited base of five volunteer users were used in order to optimize the bucket ranges of our heuristic hierarchical ranking scheme as well as the values for the parameters *tWeight, pWeight, and sWeight* for the proposed TF-scheme. The bucket ranges are as follows:

√ For the TF-heuristic, we always compute exactly 10 buckets by first computing the proposed TF metric for each publication and then we normalize the calculated scores in the range [0, 1] in a linear transformation that assigns the score 1 to the publication with the maximum calculated TF score, and then we "bucketize" the publications in the 10 intervals [0, 0.1], [0.1, 0.2], ... [0.9, 1].

√ For the 2nd-level heuristic, the bucket range is set to 5.20.

Values for the other parameters are set as follows: *sWeight* = 15.25, *pWeight* = 4.10, and *tWeight*$_{title}$ = 125.50, *tWeight*$_{abstract}$ = 45.25, and *tWeight*$_{body}$ = 5.30.

Given these parameters, we proceeded into testing the system by processing 58 *new* queries that were submitted by 15 *different users* (other than the authors of the paper) specializing in different areas of computer science and electrical and computer engineering. The users were selected based on their expertise in different areas of computer science and electrical engineering and they are researchers of different levels from the authors' universities. Each of our test users submitted a number of queries and provided feedback for all produced query results without knowing which algorithm produced each ranking. We used the three metrics mentioned before (NDCG, ERR, and LEX) to evaluate the quality of our ranking algorithm.

2.7.1 Comparisons with ACM Portal

Our ranking approach, PubSearch, compares very well with ACM Portal, and in fact outperforms ACM Portal in most query evaluations as the tests reveal using all three metrics. We illustrate the performance of each system in Table 2.1:

Table 2.1 Comparison of PubSearch with Association for Computing Machinery (ACM) Portal performance using different metrics

Metric	Number of Queries for which PubSearch Wins	Number of Queries for which ACM Portal Wins	Num. of Queries for which Both Systems Performed the Same
LEX	46	4	8
NDCG	49	1	8
ERR	44	3	11

(LEX, lexicographic ordering metric; NDCG, Normalized Discounted Cumulative Gain; ERR, Expected Reciprocal Rank).

Table 2.2 shows the average score of each system using the three different metrics:

Table 2.2 Average performance score of the different metrics

Metric	PubSearch	ACM Portal
LEX	0.742	0.453
NDCG	0.976	0.879
ERR	0.739	0.454

We witness that PubSearch performs much better than ACM Portal in most of the 58 queries used to evaluate our system under all metrics. On average, the percentage gap of performance between PubSearch and ACM Portal in terms of LEX metric is 907.5%(!), in terms of NDCG is 11.94%, and in terms of ERR, the average gap is 77.5%. The large average gap in the LEX metric is due to the fact that for some queries, ACM Portal produces a LEX score close to zero, whereas PubSearch re-orders the results so that it produces a LEX score close to 1, leading to huge percentage deviations for such queries.

Even though it is clear to the naked eye, *statistical analyses using the t-test, the sign test, and the signed rank test all show that the performance difference between the two systems is statistically significant* at the 95% confidence level for all performance metrics. In Table 2.7, we present an analytical comparison of the evaluation scores of the two systems using the three different metrics.

To highlight the difference of the ranking orders produced by the two systems, consider query#1 ("query privacy 'sensor networks' "): The ACM Portal results' list was given the following relevance judgment by the user: 1, 1, 2, 3, 2, 1, 1, 4, 3, 5. PubSearch re-orders the ACM Portal results in a sequence that corresponds to the following relevance judgment: 5, 4, 3, 3, 1, 2, 2, 1, 1, 1. PubSearch produces the best possible ordering of the given search results (with the exception of the document in the 5th position that should have been placed in the 7th position). Similarly, consider query #46 ("resource management grid computing"): ACM Portal orders its top 10 results in a sequence that received the following scores: 1, 1, 3, 3, 4, 4, 4, 5, 1, 3. PubSearch on the other hand re-orders the list of results so that the sequence's scores appear as follows: 5, 4, 3, 4, 4, 3, 1, 1, 3, 1, which is a much improved ordering than ACM Portal.

2.7.2 Comparison with other Heuristic Configurations

In Table 2.3, a head-to-head comparison of the performance of our hierarchical heuristic scheme using our custom implementation of the TF heuristic against using the traditional Boolean method is given. The results clearly show that our implementation of the heuristic outperforms the "traditional" TF heuristic.

In Figures 2.5–2.7, we show the effects of the third and last heuristic in our proposed hierarchy (using the three different metrics), namely, the ranking based on the matching of a paper's index terms to maximally

Table 2.3 Comparing the hierarchical heuristic scheme (complete, including all three levels of heuristics) using our implementation of the term frequency (TF) heuristic against the simple, Boolean TF heuristic

#	PubSearch Boolean TF LEX	PubSearch TF LEX	PubSearch Boolean TF NDCG	PubSearch TF NDCG	PubSearch Boolean TF ERR	PubSearch TF ERR
1	0.553	0.939	0.938	0.995	0.537	0.978
2	0.709	0.748	0.959	1.000	0.627	0.656
3	0.591	0.999	0.914	0.990	0.604	0.984
4	0.670	0.750	0.930	0.998	0.618	0.664
5	0.388	0.990	0.884	1.000	0.540	0.984
6	0.792	1.000	0.962	1.000	0.731	0.984
7	0.798	0.990	0.976	1.000	0.732	0.984
8	0.760	0.998	0.953	1.000	0.687	0.984
9	0.534	0.748	0.934	0.995	0.482	0.654
10	0.760	1.000	0.947	1.000	0.687	0.984
11	0.063	0.445	0.751	0.780	0.311	0.407
12	0.314	0.890	0.918	1.000	0.423	0.974
13	0.000	0.000	0.803	0.803	0.117	0.117
14	0.000	0.000	0.803	1.000	0.117	0.124
15	0.559	0.988	0.920	0.991	0.539	0.984
16	0.559	0.988	0.922	0.996	0.539	0.984
17	0.563	0.896	0.967	0.986	0.599	0.975
18	0.542	0.748	0.955	0.993	0.496	0.661
19	0.751	0.950	0.933	0.963	0.674	0.981
20	0.550	0.790	0.937	0.998	0.514	0.730
21	0.719	0.998	0.916	0.996	0.664	0.984
22	0.543	0.948	0.924	0.984	0.512	0.980
23	0.952	0.998	0.968	1.000	0.981	0.984
24	0.720	0.988	0.929	0.995	0.664	0.984
25	0.709	0.748	0.960	0.992	0.625	0.653
26	0.759	0.989	0.952	0.994	0.687	0.984
27	0.260	0.498	0.910	1.000	0.264	0.379
28	0.091	0.488	0.893	0.989	0.217	0.346
29	0.550	0.748	0.968	0.998	0.503	0.656
30	0.750	0.750	1.000	1.000	0.668	0.668
31	0.302	0.740	0.891	0.997	0.346	0.638
32	0.540	0.700	0.999	0.999	0.473	0.598
33	0.110	0.747	0.848	0.988	0.329	0.653
34	0.758	0.990	0.954	0.998	0.686	0.984
35	0.760	0.992	0.951	0.993	0.687	0.984
36	0.000	0.000	1.000	1.000	0.085	0.085

(*Continued*)

Table 2.3 Continued

#	PubSearch Boolean TF LEX	PubSearch TF LEX	PubSearch Boolean TF NDCG	PubSearch TF NDCG	PubSearch Boolean TF ERR	PubSearch TF ERR
37	0.718	0.990	0.919	0.999	0.664	0.984
38	0.300	0.540	0.886	0.955	0.340	0.480
39	0.830	0.950	0.899	1.000	0.977	0.980
40	0.000	0.000	1.000	1.000	0.085	0.085
41	0.062	0.062	0.768	0.768	0.284	0.284
42	0.720	0.998	0.915	0.989	0.664	0.984
43	0.000	0.000	1.000	1.000	0.056	0.056
44	0.800	1.000	0.974	0.998	0.732	0.984
45	0.552	0.996	0.912	0.992	0.521	0.984
46	0.551	0.942	0.936	0.984	0.521	0.980
47	0.359	0.986	0.871	0.960	0.465	0.984
48	0.558	0.942	0.942	0.982	0.538	0.980
49	0.000	0.000	1.000	1.000	0.085	0.085
50	0.000	0.000	1.000	1.000	0.085	0.085
51	0.788	0.948	0.987	0.987	0.730	0.980
52	0.550	0.742	0.970	0.992	0.504	0.651
53	0.000	0.000	1.000	1.000	0.069	0.069
54	0.796	0.956	0.973	0.973	0.732	0.982
55	0.155	0.923	0.836	0.908	0.428	0.979
56	0.958	0.990	0.981	0.998	0.982	0.984
57	0.316	0.892	0.870	0.926	0.425	0.977
58	0.960	1.000	0.975	1.000	0.982	0.984

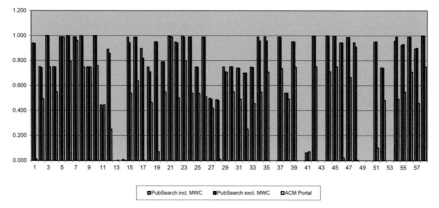

Figure 2.5 Comparison between two versions of PubSearch and Association for Computing Machinery (ACM) Portal.

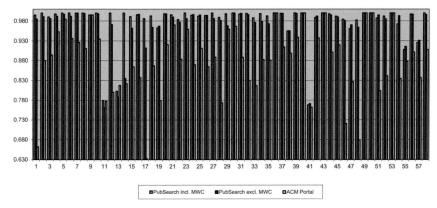

Figure 2.6 Comparison between two versions of PubSearch and ACM Portal [figure uses scores produced by the Normalized Discounted Cumulative Gain (NDCG) metric].

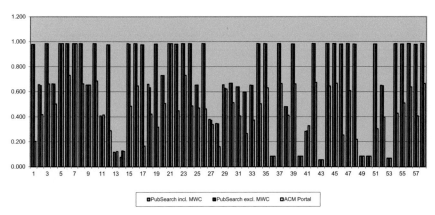

Figure 2.7 Comparison between two versions of PubSearch and ACM Portal [figure uses scores produced by the Expected Reciprocal Rank (ERR) metric].

weighted cliques in the topic graphs computed offline. The charts also show visualizations of the results in Table 2.4. Statistical analyses using the *t-test, the sign test, and the signed rank test all show that the effect of the third heuristic in the hierarchy is significant,* i.e., the hypothesis that the mean of the distribution of the percentage gap between the solutions produced by PubSearch when utilizing the 3rd heuristic in the hierarchy, and the solutions produced by PubSearch when the 3rd heuristic is excluded, is zero, must be rejected at 95% confidence level. The gap is small, but statistically significant. It is evident that all heuristics in the hierarchy are needed so as to obtain the best possible feedback score in terms of all metrics considered.

Table 2.4 Comparing TF/DCC/MWC against TF on retrieval score

Query #	TF-only LEX	TF/DCC/ MWC LEX	TF-only NDCG	TF/DCC/ MWC NDCG	TF-only ERR	TF/DCC/ MWC ERR
1	0.905	0.939	0.959	0.995	0.977	0.978
2	0.550	0.748	0.967	1.000	0.503	0.656
3	0.991	0.999	0.977	0.990	0.984	0.984
4	0.742	0.750	0.986	0.998	0.656	0.664
5	0.944	0.990	0.961	1.000	0.980	0.984
6	0.991	1.000	0.981	1.000	0.984	0.984
7	0.957	0.990	0.972	1.000	0.982	0.984
8	0.960	0.998	0.977	1.000	0.982	0.984
9	0.709	0.748	0.962	0.995	0.626	0.654
10	0.960	1.000	0.970	1.000	0.982	0.984
11	0.413	0.445	0.757	0.780	0.399	0.407
12	0.858	0.890	0.962	1.000	0.974	0.974
13	0.008	0.000	0.827	0.803	0.126	0.117
14	0.000	0.000	1.000	0.822	0.077	0.124
15	0.719	0.988	0.920	0.991	0.664	0.984
16	0.988	0.988	0.995	0.996	0.984	0.984
17	0.896	0.896	0.986	0.986	0.975	0.975
18	0.710	0.748	0.965	0.993	0.633	0.661
19	0.787	0.950	0.946	0.963	0.730	0.981
20	0.758	0.790	0.974	0.998	0.686	0.730
21	0.944	0.998	0.953	0.996	0.981	0.984
22	0.788	0.948	0.987	0.984	0.730	0.980
23	0.998	0.998	0.998	1.000	0.984	0.984
24	0.943	0.988	0.959	0.995	0.980	0.984
25	0.709	0.748	0.956	0.992	0.625	0.653
26	0.990	0.989	0.997	0.994	0.984	0.984
27	0.460	0.498	0.953	1.000	0.356	0.379
28	0.443	0.488	0.928	0.989	0.318	0.346
29	0.708	0.748	0.964	0.998	0.625	0.656
30	0.750	0.750	0.999	1.000	0.668	0.668
31	0.694	0.740	0.948	0.997	0.601	0.638
32	0.700	0.700	0.997	0.999	0.598	0.598
33	0.747	0.747	0.984	0.988	0.652	0.653
34	0.944	0.990	0.964	0.998	0.980	0.984
35	0.990	0.992	0.984	0.993	0.984	0.984
36	0.000	0.000	1.000	1.000	0.085	0.085
37	0.951	0.990	0.973	0.999	0.981	0.984
38	0.538	0.540	0.953	0.955	0.478	0.480
39	0.950	0.950	0.999	1.000	0.980	0.980

Table 2.4 Continued

Query #	TF-only LEX	TF/DCC/ MWC LEX	TF-only NDCG	TF/DCC/ MWC NDCG	TF-only ERR	TF/DCC/ MWC ERR
40	0.000	0.000	1.000	1.000	0.085	0.085
41	0.068	0.062	0.775	0.768	0.302	0.284
42	0.991	0.998	0.986	0.989	0.984	0.984
43	0.000	0.000	1.000	1.000	0.056	0.056
44	0.992	1.000	0.990	0.998	0.984	0.984
45	0.918	0.996	0.945	0.992	0.980	0.984
46	0.909	0.942	0.956	0.984	0.978	0.980
47	0.973	0.986	0.955	0.960	0.984	0.984
48	0.910	0.942	0.953	0.982	0.978	0.980
49	0.000	0.000	1.000	1.000	0.085	0.085
50	0.000	0.000	1.000	1.000	0.085	0.085
51	0.948	0.948	0.989	0.987	0.980	0.980
52	0.748	0.742	1.000	0.992	0.657	0.651
53	0.000	0.000	1.000	1.000	0.069	0.069
54	0.988	0.956	0.988	0.973	0.984	0.982
55	0.923	0.923	0.907	0.908	0.979	0.979
56	0.958	0.990	0.979	0.998	0.982	0.984
57	0.660	0.892	0.873	0.926	0.597	0.977
58	0.798	1.000	0.970	1.000	0.732	0.984

In Figures 2.8–2.10, we show the performance of our proposed heuristic configuration when comparing it with *different hierarchies of heuristics*. Each chart presents the average performance of each heuristic configuration under a different metric. Note that we specify the different heuristic hierarchies by separating each heuristic in a hierarchy with a slash ("/") character. For each hierarchy, each left-side heuristic argument is higher in the suggested hierarchy than its right-side argument.

We consider the following configurations:

1. TF/DCC/MWC (the proposed scheme)
2. TF/DCC
3. TF
4. DCC
5. MWC
6. TF/MWC

The MWC heuristic adds the most value in the full PubSearch system when measuring performance against the NDCG metric as can be seen in Figure 2.9.

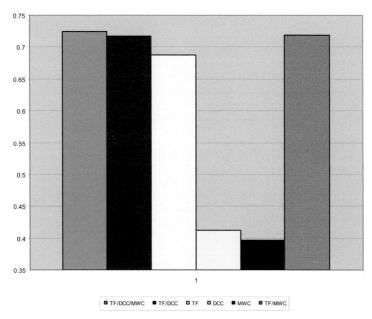

Figure 2.8   Comparison of different heuristic configurations [lexicographic ordering metric (LEX) scores].

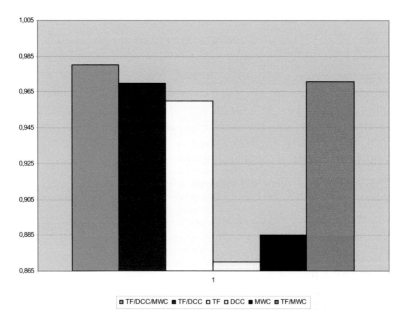

Figure 2.9   Comparison of different heuristic configurations (NDCG scores).

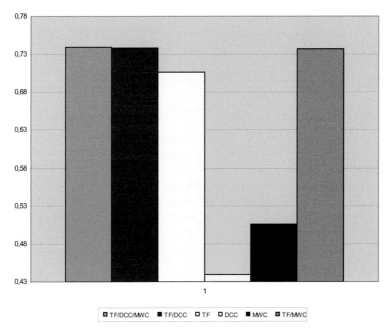

Figure 2.10 Comparison of different heuristic configurations (ERR scores).

It can be seen from these figures that our proposed configuration is the best performing configuration in terms of all metrics considered. The percentage difference between the proposed full PubSearch configuration (TF/DCC/MWC) and applying the proposed TF heuristic alone is 3.26% for the LEX metric, 1.54% for the NDCG metric, and 5.96% for the ERR metric. *Furthermore, statistical analyses using the t-test, sign test, and signed rank test show that the differences between TF/DCC/MWC and TF heuristic alone are statistically significant for all the metrics considered at the 95% confidence level.* To illustrate further, the results produced by running our proposed TF heuristic alone are shown in Table 2.4 under the multi-column labeled "TF-only," and compared against the full PubSearch system.

To make the comparison with TF-IDF methods clearer, we also compare PubSearch against the standard Okapi BM25 weighting scheme (Sparck Jones et al., 2000). However, in our comparison, since as we have already mentioned we do not maintain an academic paper database but instead simply re-rank the results returned by other engines, when computing the BM25 score for a result list, we assume that the entire database consists of the returned results of the base engine only (ACM Portal's top 10 results for a

Table 2.5   Comparing PubSearch with BM25 weighting scheme

Metric	PubSearch	Okapi BM25
LEX	0.742	0.235
NDCG	0.976	0.817
ERR	0.739	0.302

given query). As expected, the results from BM25 are quite inferior to those obtained by PubSearch. The results are summarized in Table 2.5, where we show the average score obtained for the 58 queries on each metric for each of the two systems.

The average percentage difference between PubSearch and Okapi BM25 in terms of the LEX metric is 1898% (due to BM25 producing a LEX score of less than 0.004 for some queries, while for the same query, PubSearch producing scores of more than 0.7), in terms of the NDCG metric is 20.9%, and in terms of the ERR metric, it reaches 144%. *Statistical analysis (though not really needed) in terms of t-test, sign test, and signed rank test shows these differences to be very significant.* This result is not surprising as BM25 is a generic non-binary information retrieval model that has no specific domain knowledge about academic publications.

We also compare our proposed approach against a more standard fusion scheme (Kuncheva, 2004) where for each heuristic h (that can be TF, DCC, or MWC), we compute a score A_h that represents an "inverse accuracy" score of the heuristic in obtaining the best possible sequence of a query's search results (measured against the training set query data). This score A_h is computed as follows: assume the search results for a query q ranked in descending order of relevance feedback by the user are as follows: $d_{q,1}, \ldots d_{q,n}$ having relevance feedback scores $f_{q,1} \geq f_{q,2} \geq \ldots \geq f_{q,n}$. Now, assume the heuristic h scores the documents so that they are ranked according to the following order: $d_{q,h_1}, d_{q,h_2}, \ldots d_{q,h_n}$. Define the quantity $g_{q,i}$ as follows:

$$
g_{q,i} = \begin{cases} 0 & \text{if } f_{q,i} = f_{q,h_i} \\ \left\{ \min |i - \bar{j}|, |i - \underline{j}| \right\} : \bar{j} = \max \left\{ j | f_{q,j} = f_{q,h_i} \right\}, \\ \underline{j} = \min \left\{ j | f_{q,j} = f_{q,h_i} \right\} & \text{else} \end{cases}
$$

We define $A_{q,h} = \sum_{i=1}^{n} g_{q,i}$. Clearly, $A_{q,h} \geq 0$, and $A_{q,h} = 0$ if and only if the heuristic h obtains a perfect sorting of the result set of query q (as indicated by the user relevance judgments.) The measured inverse accuracy of a heuristic h on the training set Q_{train} is then defined as $A_h = \sum_{q \in Q_{\text{train}}} A_{q,h}$. We compare PubSearch against an ensemble of the three heuristics from the

Table 2.6 Comparing PubSearch with heuristic ensemble fusion performance average

Metric	PubSearch	TF-DCC-MWC Ensemble Fusion
LEX	0.742	0.431
NDCG	0.976	0.879
ERR	0.739	0.451

set $H = \{TF, DCC, MWC\}$ that works as follows: each heuristic in the set of heuristics produces a re-ranked order of results $d_{q,h_1}, d_{q,h_2}, \ldots d_{q,h_n}$. The final ensemble result is the list of results sorted in ascending order of the combined value

$$r_d = \left[\sum_{h \in H} (A_h + 1)^{-1} \right]^{-1} \sum_{h \in H} \frac{r_{d,h}}{(A_h + 1)} \qquad (2.7)$$

of each document in the result list where $r_{d,h}$ is the position of document d in the result list according to heuristic h. The ensemble fusion results are comparable with ACM Portal on the NDCG and ERR metrics (0.5% better in terms of NDCG and 8.1% better in terms of ERR metric); the ensemble fusion also does a much better job than ACM Portal in terms of the LEX metric (442% better). Still, the ensemble fusion results do not compare well with PubSearch as can be seen in Table 2.6.

PubSearch is on average more than 472% better than the fusion heuristic described above in terms of LEX metric, more than 12% better in terms of the NDCG metric, and more than 70% better than the fusion heuristic in terms of the ERR metric.

2.7.3 Comparison with Other Academic Search Engines

We performed a head-to-head comparison between PubSearch and the three state-of-the-art academic search engines:

1. Google Scholar (http://scholar.google.com)
2. Microsoft Academic Search (http://academic.research.microsoft.com)
3. ArnetMiner (http://arnetminer.org)

The comparison was made on a sizeable subset of our original query set of 58 user queries shown in Table 2.7, comprising a total of 20 user queries, augmented by 4 new user queries, for a total of 24 user queries. The four new user queries were Q_{59} = "Page Rank clustering," Q_{60} = "social network information retrieval," Q_{61} = "unsupervised learning," and Q_{62} = "web mining," respectively.

Table 2.7   Comparing PubSearch with ACM Portal on retrieval score

#	Submitted Query	ACM LEX	PubSearch LEX	ACM NDCG	PubSearch NDCG	ACM ERR	PubSearch ERR
1	Query privacy "sensor networks"	0.012	0.939	0.662	0.995	0.204	0.978
2	Wormhole attacks adhoc networks	0.492	0.748	0.880	1.000	0.417	0.656
3	Gameplay artificial intelligence	0.748	0.999	0.894	0.990	0.663	0.984
4	Human-level AI	0.548	0.750	0.953	0.998	0.502	0.664
5	Ambient intelligence	0.988	0.990	0.985	1.000	0.984	0.984
6	Cloud computing	0.794	1.000	0.936	1.000	0.732	0.984
7	Autonomous agents	0.961	0.990	0.926	1.000	0.984	0.984
8	Service-oriented architecture	0.748	0.998	0.911	1.000	0.664	0.984
9	Routing wavelength assignment heuristic	0.748	0.748	0.995	0.995	0.654	0.654
10	GMPLS "path computation"	0.760	1.000	0.935	1.000	0.687	0.984
11	Background subtraction for "rotating camera"	0.446	0.445	0.777	0.780	0.415	0.407
12	Image registration in "video sequences"	0.250	0.890	0.800	1.000	0.290	0.974
13	Computer vision code in MATLAB	0.002	0.000	0.817	0.803	0.123	0.117
14	Secure decentralized voting	0.002	0.000	0.822	1.000	0.124	0.124
15	License plate recognition	0.539	0.988	0.864	0.991	0.485	0.984
16	Ellipse fitting	0.638	0.988	0.837	0.996	0.647	0.984
17	Single channel echo cancellation	0.000	0.896	0.632	0.986	0.166	0.975
18	Analysis time-varying systems	0.466	0.748	0.867	0.993	0.422	0.661
19	Time-varying system identification	0.070	0.950	0.779	0.963	0.318	0.981
20	Amazon mechanical turk	0.548	0.790	0.921	0.998	0.506	0.730

Table 2.7 Continued

#	Submitted Query	ACM LEX	PubSearch LEX	ACM NDCG	PubSearch NDCG	ACM ERR	PubSearch ERR
21	Music color association	0.988	0.998	0.970	0.996	0.984	0.984
22	Mobile TV user experience	0.502	0.948	0.883	0.984	0.447	0.980
23	Mobile television convergence	0.796	0.998	0.959	1.000	0.732	0.984
24	Music instrument recognition	0.540	0.988	0.869	0.995	0.487	0.984
25	Bayesian n-gram estimation prior	0.536	0.748	0.911	0.992	0.470	0.653
26	Statistical parametric speech synthesis	0.510	0.989	0.864	0.994	0.462	0.984
27	Cover song identification	0.419	0.498	0.889	1.000	0.339	0.379
28	Bayesian spectral estimation	0.012	0.488	0.773	0.989	0.161	0.346
29	Object-oriented programming	0.708	0.748	0.959	0.998	0.623	0.656
30	XML database integration	0.550	0.750	0.966	1.000	0.512	0.668
31	Agile software development	0.491	0.740	0.889	0.997	0.406	0.638
32	Script languages	0.251	0.700	0.829	0.999	0.266	0.598
33	Distributed computing web services	0.458	0.747	0.817	0.988	0.373	0.653
34	Database performance tuning	0.548	0.990	0.882	0.998	0.503	0.984
35	Database scaling	0.708	0.992	0.881	0.993	0.631	0.984
36	Database optimization	0.000	0.000	1.000	1.000	0.085	0.085
37	Distributed database architecture	0.735	0.990	0.915	0.999	0.665	0.984
38	Large-scale database clustering	0.492	0.540	0.899	0.955	0.412	0.480
39	Autonomous agents and multi-agent systems	0.747	0.950	0.939	1.000	0.663	0.980

(*Continued*)

Table 2.7   Continued

#	Submitted Query	ACM LEX	PubSearch LEX	ACM NDCG	PubSearch NDCG	ACM ERR	PubSearch ERR
40	Distributed autonomous agents	0.000	0.000	1.000	1.000	0.085	0.085
41	Self-organizing autonomous agents	0.070	0.062	0.762	0.768	0.329	0.284
42	Large-scale distributed middleware	0.752	0.998	0.937	0.989	0.675	0.984
43	Intelligent autonomous agents	0.000	0.000	1.000	1.000	0.056	0.056
44	Grid computing cloud computing	0.712	1.000	0.901	0.998	0.646	0.984
45	Cloud computing platforms	0.750	0.996	0.920	0.992	0.668	0.984
46	Resource management grid computing	0.020	0.942	0.721	0.984	0.254	0.980
47	Cloud computing architectures	0.666	0.986	0.827	0.960	0.610	0.984
48	Cloud computing state of the art	0.004	0.942	0.679	0.982	0.218	0.980
49	User interface technologies	0.000	0.000	1.000	1.000	0.085	0.085
50	Mobile user interfaces	0.000	0.000	1.000	1.000	0.085	0.085
51	Web 2.0	0.100	0.948	0.804	0.987	0.304	0.980
52	Mobile social networks	0.483	0.742	0.841	0.992	0.397	0.651
53	Social network privacy	0.000	0.000	1.000	1.000	0.069	0.069
54	Game engine architecture	0.492	0.956	0.835	0.973	0.428	0.982
55	3D game engine	0.549	0.923	0.878	0.908	0.509	0.979
56	OpenGL	0.709	0.990	0.902	0.998	0.638	0.984
57	Texture mapping	0.460	0.892	0.837	0.926	0.405	0.977
58	Polygonal meshes	0.748	1.000	0.908	1.000	0.666	0.984

Each query was given to each of the above-mentioned search engines, and the top-10 results (from each of the above search engines) were then presented to the users for relevance feedback in random order. The results produced by each search engine, as well as the re-ranked results produced by

Table 2.8 Comparison between Microsoft Academic Search and PubSearch

Query #	Microsoft Academic Search LEX	PubSearch LEX	Microsoft Academic Search NDCG	PubSearch NDCG	Microsoft Academic Search ERR	PubSearch ERR
59	0.52	0.72	0.89	0.91	0.5	0.66
60	0.47	0.86	0.8	0.85	0.48	0.97
61	0.88	0.80	0.89	0.97	0.97	0.73
62	0.99	0.96	0.97	0.96	0.98	0.98
34	0.74	0.71	0.97	0.95	0.64	0.62
35	0.45	0.49	0.85	0.89	0.38	0.41
36	0.49	0.46	0.98	0.96	0.35	0.34
37	0.69	0.75	0.92	0.99	0.61	0.65
58	0.44	0.46	0.9	0.94	0.33	0.35
38	0.67	0.79	0.89	0.99	0.62	0.73
39	0.70	0.53	0.99	0.98	0.57	0.44
40	0.46	0.45	0.92	0.92	0.35	0.35
41	0.30	0.69	0.9	0.95	0.31	0.56
42	0.50	0.46	0.98	0.93	0.38	0.36
43	0.24	0.25	0.91	0.93	0.21	0.22
44	0.45	0.46	0.89	0.91	0.37	0.39
45	0.29	0.29	0.91	0.93	0.27	0.28
46	0.74	0.71	0.94	0.94	0.63	0.61
47	0.31	0.53	0.91	0.97	0.34	0.45
49	0.46	0.49	0.95	0.98	0.33	0.35
54	0.71	0.74	0.95	0.98	0.62	0.64
55	0.53	0.46	0.96	0.9	0.45	0.38
56	0.25	0.24	0.97	0.95	0.18	0.18
57	0.48	0.77	0.87	0.98	0.48	0.72

PubSearch when given the same list of results are shown in Tables 2.8–2.10, respectively.

The average percentage difference between PubSearch and Microsoft Academic Search is 15% for the LEX metric, 2.6% for the ERR metric, and 11.7% for the NDCG metric. *Statistical analysis shows that for the ERR and LEX metric, the differences are significant at the 95% confidence level according to the t-test and signed-rank test but not according to the sign test.* The results are visualized in Figure 2.11.

The average percentage difference between PubSearch and Google Scholar in terms of LEX is 39%, in terms of ERR metric is 4.6%, and in terms of NDCG is 13.4%. The results of applying the t-test, the signed rank test, as well as the sign test *on the ERR metric* shows that

Table 2.9 Comparison between Google Scholar and PubSearch

Query #	Google Scholar LEX	PubSearch LEX	Google Scholar NDCG	PubSearch NDCG	Google Scholar ERR	PubSearch ERR
59	0.45	0.56	0.4	0.53	0.79	0.91
60	0.34	0.71	0.41	0.64	0.84	0.9
61	0.99	1.00	0.98	0.98	0.95	0.98
62	0.19	0.99	0.51	0.98	0.83	0.97
34	0.94	0.79	0.97	0.72	0.98	0.99
35	0.48	0.54	0.393	0.44	0.91	0.98
36	0.31	0.66	0.345	0.57	0.91	0.94
37	0.61	0.54	0.54	0.46	0.87	0.99
58	0.96	0.76	0.98	0.68	0.97	0.94
38	0.51	0.55	0.446	0.5	0.9	0.96
39	0.66	0.53	0.54	0.43	0.95	0.99
40	0.27	0.46	0.307	0.38	0.86	0.91
41	0.62	0.53	0.54	0.44	0.89	0.96
42	0.75	0.70	0.65	0.62	0.99	0.95
43	0.91	0.99	0.97	0.98	0.91	0.99
44	0.69	0.53	0.6	0.48	0.91	0.91
45	0.71	0.74	0.62	0.64	0.96	0.98
46	0.78	0.98	0.73	0.98	0.96	0.98
47	0.70	0.50	0.59	0.42	0.99	0.95
49	0.99	0.70	0.98	0.62	0.99	0.89
54	0.70	0.53	0.57	0.44	0.99	0.98
55	0.95	0.76	0.98	0.68	0.98	0.97
56	0.31	0.94	0.376	0.97	0.85	0.98
57	0.31	0.78	0.39	0.72	0.87	0.99

the improvement is statistically significant at the 95% confidence level. However, the same does *not* apply for the other two metrics, although the *t*-test shows that the results for the LEX metric are also statistically significant at the 93% confidence level. A visualization of the comparison results is shown in Figure 2.12.

The average percentage improvement of PubSearch over the ArnetMiner results in terms of LEX score is 19.9%, in terms of ERR is 4.1%, and in terms of NDCG metric is 12.6%.

The results of *all* statistical tests are statistically significant for the LEX and ERR metrics, but only the sign test shows statistical significance for the NDCG metric at the 95% confidence level. Figure 2.13 visualizes these results.

Table 2.10 Comparison between ArnetMiner and PubSearch

Query #	ArnetMiner LEX	PubSearch LEX	ArnetMiner NDCG	PubSearch NDCG	ArnetMiner ERR	PubSearch ERR
59	0.56	0.72	0.88	0.94	0.6	0.66
60	0.27	0.75	0.8	0.97	0.35	0.68
61	0.36	0.80	0.87	0.97	0.46	0.73
62	0.80	0.95	0.96	0.96	0.73	0.98
34	0.95	0.76	0.97	0.95	0.98	0.68
35	0.75	0.75	0.99	0.99	0.66	0.66
36	0.50	0.51	0.9	0.91	0.43	0.45
37	0.55	0.59	0.91	0.93	0.51	0.6
58	0.73	0.69	0.95	0.92	0.63	0.6
38	0.53	0.55	0.85	0.89	0.49	0.53
39	0.45	0.46	0.9	0.93	0.34	0.36
40	0.46	0.74	0.88	0.98	0.41	0.65
41	0.70	0.54	0.87	0.89	0.61	0.51
42	0.71	0.74	0.9	0.94	0.62	0.65
43	0.49	0.49	0.84	0.86	0.41	0.43
44	0.55	0.71	0.93	0.94	0.49	0.62
45	0.59	0.79	0.94	0.98	0.6	0.73
46	0.46	0.49	0.91	0.98	0.33	0.36
47	0.70	0.70	0.89	0.91	0.61	0.62
49	0.69	0.54	0.94	0.96	0.6	0.48
54	0.67	0.69	0.92	0.93	0.6	0.6
55	0.30	0.49	0.88	0.92	0.31	0.39
56	0.70	0.51	0.97	0.96	0.58	0.43
57	0.65	0.73	0.88	0.98	0.55	0.62

2.7.4 Can PubSearch Promote Good Publications "Buried" in ACM Portal Results?

For the query "Web Information Retrieval," because ACM Portal fails to return "The anatomy of a large-scale hyper textual web search engine" paper within the first 5,000 search results, we manually added the "Google" paper into the top 50 results from ACM Portal and asked PubSearch to re-rank the new list of 51 search results. The Page-Rank paper comes out in the 5th place, immediately below the following papers:

1. Contextual relevance feedback in web information retrieval (Limbu et al., 2006).
2. Concept unification of terms in different languages via web mining for information retrieval (Li et al., 2009).

3. An architecture for personal semantic web information retrieval system (Yu et al., 2005).
4. An algebraic multi-grid solution of large hierarchical Markovian models arising in web information retrieval (Krieger, 2011).

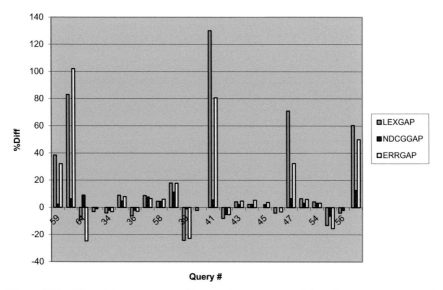

Figure 2.11 Plot of the percentage difference between the PubSearch score and Microsoft Academic Search score in terms of the three metrics LEX, ERR, and NDGC.

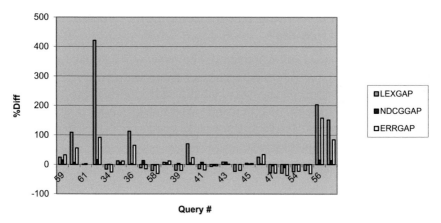

Figure 2.12 Plot of the percentage difference between the PubSearch and Google Scholar.

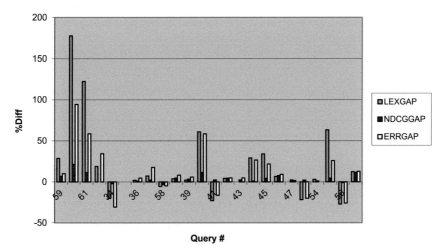

Figure 2.13   Plot of the percentage difference between the PubSearch and ArnetMiner.

The papers appearing above the Page-Rank paper all share the following characteristics: (i) they have all terms of the query appearing in the title and (ii) they are more recent papers. Because of this, our custom implementation of the TF heuristic promotes the other papers high in the result list so that the Google paper ends up in the 2nd TF bucket, and then, its citation count alone cannot promote it higher than the 5th position. Still, PubSearch manages to promote the Google paper in the top five results which is much better than the other academic search engines we experimented with.

To further enhance our confidence in the ability of PubSearch to promote "good" publications – for a particular user information need – that happen to appear much lower than the top 10 positions in the results' list of ACM Portal, we ran a small experiment where the users ranked the top 25 results of 5 queries. The results are very good, as the system shows again very significant performance improvement against ACM Portal in all metrics considered, and in fact, it significantly *improves* its performance gap over ACM Portal in terms of both the LEX and the NDCG metrics.

The results are shown in Table 2.11. The percentage improvement of Pub-Search over ACM Portal on average in terms of the LEX metric is 460.98%, in terms of the ERR metric is 16.7%, and in terms of the NDCG metric is 111.6%. All the results are statistically significant at the 95% confidence level. The very high gap in terms of LEX score for the case of the top-25 results is exactly due to the fact that good publications that are a good

Table 2.11   Limited comparison between ACM Portal and PubSearch for the top-25 results of ACM Portal. Q63 is the query "clustering 'information retrieval' "

Query #	ACM LEX	PubSearch LEX	ACM NDCG	PubSearch NDCG	ACM ERR	PubSearch ERR
33	0.46	1.00	0.79	0.99	0.38	0.98
34	0.55	0.99	0.87	0.99	0.50	0.98
35	0.70	0.95	0.90	0.97	0.63	0.98
36	0.00	0.69	0.68	0.98	0.12	0.57
37	1.00	0.99	0.98	0.98	0.98	0.98
38	0.45	1.00	0.84	0.98	0.42	0.98
39	0.75	0.95	0.91	0.99	0.66	0.98
40	0.00	0.49	0.72	0.99	0.11	0.36
41	0.07	0.99	0.79	0.96	0.33	0.98
42	0.75	0.99	0.92	0.97	0.68	0.98
45	0.75	1.00	0.90	0.99	0.67	0.98
46	0.02	0.80	0.75	0.97	0.25	0.73
59	0.46	0.98	0.83	0.97	0.44	0.98
60	0.04	0.78	0.75	0.97	0.22	0.73
61	0.88	0.95	0.87	0.96	0.97	0.98
62	0.31	0.96	0.86	0.97	0.42	0.98
63	0.33	0.74	0.86	0.98	0.39	0.66

match for the user's information needs are actually promoted from the bottom of the list of the top-25 ACM Portal results to the top positions. The LEX score is therefore a useful indicator when investigating the ability of ranking schemes to promote otherwise "buried" publications high in the result list as it amplifies this effect to the maximum extent.

2.7.5 Run-time Overhead

The run-time overhead of our *initial prototype* requires to perform the re-ranking of the search results given a query and a set of results from another engine (i.e., ACM Portal, or Google Scholar or Microsoft Academic Search, or ArnetMiner) is in the order of 3 *seconds per document. This run-time applies for a commodity hardware workstation.* However, the computation of the TF-score (by far the most compute intensive process in the whole system) for each document is independent of the other documents in the result list, and therefore can be done in parallel so that the total computation time for a full list of search results will still be in the order of seconds in a server farm. Furthermore, our prototype is not in any way optimized for speed at the moment of writing.

3

Recommender Systems

3.1 System Architecture Overview

A commercial movie recommendation system (called AMORE) has been developed for a major Greek Triple Play services provider. The provider uses the Microsoft Media Room® movie rental platform that allows service subscribers to stream movies online. AMORE is implemented based on a service-oriented architecture (SOA) which aims to expose only the required interfaces to service consumers, without revealing any implementation details. Similarly, AMORE retrieves all updated data, related to user transactions, as well as available content items, via exposed *data retrieval* web services from the provider's side. This loose-coupling of the overall architecture design allows for flexible integration of all involved subsystems.

Furthermore, AMORE, in addition to offering *on-the-fly* generation of recommendations, also supports a daily update and caching of recommendations in an attempt to minimize computational overhead in the deployment configuration which has limited resources. In order to achieve that, AMORE is divided into the following components: (i) the AMORE web service, (ii) the AMORE batch job process, and (iii) two different database instances that comply with the exact same data model. The AMORE batch job generates and caches a predefined number of recommendations for each active subscriber of the service, as well as overall top-n recommendations (for all users), both of which facilitate the AMORE web service methods to retrieve cached recommendations with the minimum number of operations.

In the next chapter, we will describe the architecture of the AMORE system as well as the design of the recommendation algorithms used. We have not been involved in any technical aspect of the movie rental platform itself, which is entirely operated by the triple play provider, so anything related to that system is considered irrelevant to the current work and will be skipped.

3.1.1 AMORE Web Service

The AMORE web service exposes a set of recommendation related methods. These methods can be distinguished into: (i) those that generate recommendations *on-the-fly* by processing user histories real-time using different *views* of the database, bounded by specific time limits, and (ii) those that simply retrieve daily updated cached recommendations from DB.

Caching of recommendations speeds up system responsiveness to web service client requests by eliminating the overhead of having the recommender engine generate recommendations on-the-fly. Having access to cached results, the web service can easily retrieve recommendations with the minimum number of operations [at standard time $O(1)$, by means of a simple SQL *select* statement from a single table] to retrieve a specified number of recommendations for a specific user. Also due to the fact that there are no significant variations in the watching behavior of a single user account within very small intervals of time (within 1 or 2-hour units), caching of the recommendations proves to be justified, and it is a business decision (which does not relate to the current discussion) on the frequency with which recommendations are going to be updated at the movie rental platform, although it is worth noting that we are able to *refresh* results every 2 hour approximately for a database containing nearly a year's worth of transactions from approximately 30,000 users.

In addition to caching of results, there are service methods where caching does not cannot apply; these are methods that offer service consumers the possibility to narrow down the user transaction history to a specific *subset* bound by specific time limits based on which recommendations are generated. This functionality allows service consumers to retrieve recommendations based on user histories of different *time slots* within the day (we define as a time slot a 3-hour interval, but the range is fully configurable) with the aim to push different recommendations corresponding to each (or even a sequential combination of more than one) time slot. Most likely, different kinds of users (bound with a single account) have different watching behaviors at different time slots within a day, so to illustrate with a rather simplified example, most likely, we would expect young children to watch movies during morning hours, elderly people during early evening hours, and adults during late evening hours. So, providing time-bound recommendations allows the system to address preferences of users who happen to watch movies at specific moments within a day mapped with certain time slots.

When the user logs into the triple play provider's front-end screen, he is presented with personal, customized account-based recommendations, essentially retrieved by calling the respective web service method. As we will describe in the following section, the caching of results in DB is performed by a batch job that performs a series of tasks for the "offline" updating and caching of the recommendation results. To facilitate the uninterrupted running of both – the web service and the batch job in parallel – the system uses two databases (schemas) that we will refer to as the *main* and the *auxiliary* that follow the exact same data model. Having two schemas allows our system to serve web service requests by retrieving cached results from one schema (for instance, the *main*) while the batch job uses the second schema (in this case, the *auxiliary*) to store the updated results.

3.1.2 AMORE Batch Process

Since AMORE is configured to run in a Linux-based OS, the batch process has been designed to run as a CRON job. The CRON job is registered to run *indefinitely* at specified time intervals attempting to spawn a new instance of the batch process, in case one is not already running. The batch process runs a series of steps in order to update and cache new recommendations that the exposed web service methods can retrieve upon completion of all steps. As already mentioned, the overall system architecture includes the existence of two schemas (we will reference them as *main* and *auxiliary*) that allow the uninterrupted, efficient service of web service requests, while at the same time, updated recommendations are generated as part of a background process. The batch process also controls the data source references used by each component and is also responsible for updating them (by switching pointers) upon process completion, as we will explain in detail later.

The AMORE batch process runs a series of steps sequentially. First, the process determines the data source reference to use, which has to be the *opposite* of the data source currently used by the web service. To illustrate, say that the web service uses the *main* data source reference, retrieving cached recommendations from the *main* database, the batch process should then establish a DB connection using the data source reference to the *auxiliary* schema, which must contain the most outdated recommendations, since the web service always serves recommendation requests retrieving data from the database containing the most recent recommendations.

The first real task that the batch process performs is attempting to verify that all external, back-end web services are *responsive* to submitted web service requests, the AMORE batch job calls two provision web services (*user* and *data* respectively), in order to update the system's database with the latest data for all active users, and that includes all user transactions as well as the full list of available content items. After all retrieved are stored in persistence, the system refreshes all memory caches (*user* and *item*, respectively) that will be referenced by recommenders in the recommendation generation step. So, after caches have been loaded and renewed, the process invokes the recommender ensemble (described in-depth in the next section) in order to efficiently generate for each active user the updated top-*n* recommendations and cache them, so that they can be later retrieved by the web service upon process completion. A certain pre-specified number of recommendations are generated of size that is considered to be sufficient in order to cover all different business requirements of the provider. The provider, at a middle tier in the overall architecture, may perform some filtering on the generated recommendation results by promoting at higher positions certain items which they consider to be *trending*, or others that are deemed of high relevance to a specific user (by the recommender engine) and are priced higher than other items in the list, which might occupy higher positions. Also there is a business case where the provider might be even willing to eliminate certain items from the list (or move them to lower positions in the rank), because they are lower (or even zero-) priced. But this discussion is beyond the scope of the current work, and does not affect in any way the experiments that we will present.

One of the frequently occurring issues that our system needed to address concerns recommendations that keep on appearing for a certain user (since the items are considered to be relevant), but these items are never consumed by the user. Some of the main reasons that may cause this are the fact that the user may have seen the recommended item in the past via a different broadcasting medium, the fact that a specific recommendation is not attractive enough, and even highly priced based on the maximum reference price that a specific user is willing to pay. So, in order to avoid a situation where certain recommendations remain *forever* in the suggested recommendations of a specific user, without being eventually consumed, we have devised a post-processing mechanism that discounts the score assigned by our recommender ensemble for each item, proportionally to the number of times that the specific item has been shown in the recommendation list of a specific user. This would eventually cause recommendations that are shown, but are not consumed, to eventually "fall off the charts" [see Lathia et al. (2010) for a detailed evaluation of methods for solving exactly this kind of *temporal diversity issue in recommender systems*; (Hurley and Zhang, 2011) for a thorough review of different approaches to the related problem of maximizing diversity to item recommendations to users; Zhang and Hurley (2008) formulate an optimization problem for maximizing diversity in recommendation lists subject to maintaining high relevance of the recommended items].

One additional frequent issue for recommender systems is the cold start problem which refers to the inability of a recommender system to generate accurate recommendations for a specific user due to lack of user data by means of transaction history. This is a frequent issue in cases of *new* users to the service. Also the problem persists but takes a different form, in cases of users with very small histories. This would still allow recommenders to come up with recommendation results; still recommendations cannot be accurate enough until the user builds a transaction history that would allow the system to generate recommendations with higher confidence. In order to address the first part of the cold start problem which refers to new users with no transaction histories whatsoever, the batch process generates a list of top-*n* recommendations, which corresponds to a sorted list of all items in the DB, ordered using the lexicographic ordering rule. Specifically, the ordering happens by generating a vector of integer values v_{it} for each item in the database of size *n* corresponding to the total number of recommendations generated for each user at the previous step of the process. Each element of v_{it} contains the count value corresponding to the number of times that the particular item *it* appeared in the i_{th} position in the top-*n* result list of each user for which recommendations have been generated.

As a final step, after all recommendations have been generated and successfully cached, the batch process calls a web service method (only available to the process) that instructs the AMORE web service to switch the active data source from the currently used schema to its opposite; after doing so, the web service is able, upon request, to return cached recommendation of results stored in the schema that the last process instance has populated. Similarly, the AMORE job updates a file that specifies to the next process instance that will be spawned at some point by the CRON scheduler to use a data source reference to the schema containing the most outdated results in order to repeat the aforementioned process.

Furthermore, in order to ensure that all cached data as well as all data retrieved from persistence at any point are consistent with the currently active data source, we have implemented a mechanism that ensures and protects the system resources from "dirty" reads/writes. This mechanism is based on a *fast reentrant global read/write lock* with the following properties:

1. A thread owning the write-lock may request (and get) the same lock in read- or write-mode any number of times, but must call the corresponding release method for every time it has called the request method in order for the locks to be eventually released.
2. A thread owning a read-lock may request an upgrade to the write-lock, and the method will grant the new type lock, unless at the time of request is at least one other thread having the read-lock, in which case there is a danger of dead-lock; in such a situation, a checked exception is thrown.
3. Threads executing a request for a read-lock will yield the *first* time if there exists a thread waiting for the write-lock so as to avoid any possible live-lock issues.

Given this global lock, we implement a simple pattern in all related methods for creating, maintaining, and/or updating the in-memory caches: whenever a method needs to access the in-memory caches, it must first obtain the global read-lock, whereas methods that need to update the in-memory caches must first obtain the global write-lock. Upon start-up of the AMORE web-service, the first thread started spawns a new thread that obtains the global write-lock and starts loading the data from the database into the in-memory caches, while the first thread waits for the new thread to complete [calling the thread's *join*() method]. Once the new thread has loaded the latest snapshot of the database, it releases the write-lock and finishes, returning control to the first thread to continue its operation. Coordination between the AMORE batch job and the AMORE web service (two distinct processes residing in distinct address spaces) is obtained as follows: when the AMORE batch job is about to complete, as a last step, it calls the special AMORE web service method mentioned above, which in turn, first obtains the global write-lock of the system, then switches the DB pointer to the current active DB schema, then refreshes all in-memory caches of the system, and finally releases the global write-lock, allowing pending recommendation requests (waiting to obtain the global read-lock) to proceed using the most updated data. Figure 3.1 provides a visual representation of the overall system architecture, as discussed above.

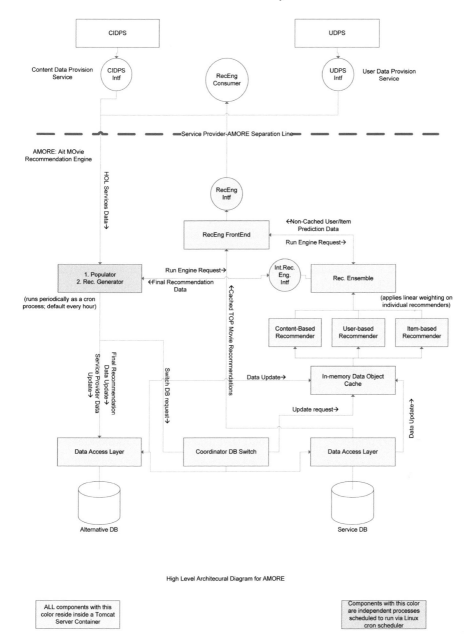

High Level Architecural Diagram for AMORE

ALL components with this color reside inside a Tomcat Server Container

Components with this color are independent processes scheduled to run via Linux cron scheduler

Figure 3.1 AMORE: High-level architecture.

3.2 Recommender Ensemble

3.2.1 Recommendation Approach

In configurations where no relevance judgments exist, there have been different attempts to determine user preference by means of different available parameters. Specifically, Christou et al. (2012) introduce a novel content-based recommender that takes a pure machine learning approach for performing movie recommendation functionality to subscribers of a content-delivery network. The introduced approach uses the percentage of watching time of a specific show by a specific user as a criterion that determines the degree of preference that the user has for the watched show. Based on that knowledge, the introduced recommender attempts to classify a certain movie into one of two classes ("like" or "dislike"). A similar idea has been used in Bambini et al. (2011), where the authors introduce an online classifier ensemble based on the Hedge-β algorithm to determine class membership of previously unseen content in order to be able to recommend shows in the category "*like.*"

The aforementioned approach could not be applied in our own case, since there has been no such information with respect to a particular user's watching behavior or the total watching time devoted to a certain movie item. Additionally, the complete absence of any user relevance judgments or feedback led us to resort on an algorithmic approach that uses a user's transaction history, in terms of consumed content items, and based on that information alone, we designed and implemented a hybrid recommender ensemble composed of a (i) *content-*, (ii) *item-*, and (iii) *user*-based recommender.

3.2.2 Content-Based Recommender

To explain the algorithm behind the content-based recommender we implemented, we define set $H_u = \{p_{u,i_1}, \ldots, p_{u,i_k}\}$ to denote the set of item content item purchases as evidenced in the transaction history of user u.

For each purchase $p_{u,i}$ in the set H, we have the following associated meta-data:

1. An ordered list of actors A_i that appear in the content-item i (in order of appearance). Each element in this list is an element of the full set of actors A known to the system.
2. An ordered list of directors D_i that directed the content-item. Each element in this list is an element of the full set of directors D known to the system.

3. An ordered list of producers P_i that produced the content-item. Similarly, each element of the list is a member of the set P of producers known to the system.
4. An ordered list of the genres G_i of the content-item, each element of which is a member of the full set of genres G that the service provider has defined.
5. The year y_i the content-item was produced.
6. An ordered list of the countries C_i that participated in the production of the content-item.
7. An ordered list of the languages L_i in which the content-item is available.
8. An ordered list of the languages S_i in which subtitles in the content-item is available.
9. The total duration of the content-item d_i (in seconds).
10. The price $m_{u,i} \geq 0$ the user paid to view the item.
11. The exact date and time $t_{u,i}$ the user started viewing the content-item.

Given the above information, our custom content-based recommender is able to compute the following functions:

$$F^A(x,u) = \sum_{i:x \in A_i \wedge p_{u,i} \in H_u} [1 + m_{u,i}]$$

$$F^D(x,u) = \sum_{i:x \in D_i \wedge p_{u,i} \in H_u} [1 + m_{u,i}]$$

$$F^P(x,u) = \sum_{i:x \in P_i \wedge p_{u,i} \in H_u} [1 + m_{u,i}] \tag{3.1}$$

$$F^G(x,u) = \sum_{i:x \in G_i \wedge p_{u,i} \in H_u} [1 + m_{u,i}]$$

$$F^Y(x,u) = \sum_{i:|x-y_i|<l^y \wedge p_{u,i} \in H_u} [1 + m_{u,i}]$$

Similarly, the functions F^L, F^S, F^C are defined; all are cached in appropriate hash-tables in memory so that the computations are only performed once, right after the system's databases are updated. Each of the above functions provides an estimate of the degree of "matching" of a user u with the value of the appropriate attribute x: for example, F^A ("Tom Cruise," "S668275") represents the system's estimate of the matching of user "S668275" with actor "Tom Cruise," and the estimate is essentially the sum of euro the user has paid to see movies starring Tom Cruise plus the total number of times the user saw movies starring that actor.

The prediction score of a content item i that has not been already viewed by user u then is computed according to the following formula:

$$R_{u,i} = \sum_{j \in \{A,D,P,G,L,S,C,Y\}} w^j R^j_{u,i}/M_u \tag{3.2}$$

where the quantities $R^j_{u,i}$, M_u are defined as follows:

$$M_u = \sum_{j \in \{A,D,P,G,L,S,C,Y\}} w^j \cdot \sum_{i:p_{u,i} \in H_u} |j_i| \cdot \left[\max_x \left\{ F^j(x,u) : x \in j_i \right\} \right]^{k^j}$$

$$R^j_{u,i} = w^j \cdot \sum_{x \in j_i} \left[F^j(x,u) \right]^{k^j} \tag{3.3}$$

and the set Y_i is defined as $Y_i = \{x \in \mathbb{N}:|x - y_i| < l^y\}$ where l^y is a non-negative parameter. The score $R_{u,i} \in [0,1]$ is therefore a weighted non-linear combination of the "likeness" of the user toward each of the content-item attributes as measured by the total percentage of the amount of money the user has paid to view items with such an attribute as well as by the number of times the user has viewed such items. The top-*n* recommendations are the *n* items currently available for viewing having the highest $R_{u,i}$ score for each user u.

The parameters l^y, w^j as well as the exponents k^j for $j = A, D, P, G, L, S, C, Y$ were considered to be independent non-negative variables to be optimized, with objective criterion the recall metric $R(10)$ specified in the experimental results section. Different values for the l^y, w^j, k^j produce different recall metric values. We optimized these parameters, using again the popt4jlib Open Source library via a standard genetic algorithm process.

3.2.3 Item-Based Recommender

Our custom implementation of the *k-NN-item-based recommender* is as follows (we simplify somewhat our description to avoid discussing issues that are not essential to the algorithm such as availability of content-items, filtering of the user histories according to certain time-windows, etc.).

Let U denote the set of all users that have subscribed at some point to the video-on-demand service; for every user $u \in U$, let their unique sequential user-id be $sid(u) \in \{1, \ldots |U|\}$, and similarly let I be the set of all content-items known to the system, and for every item, $i \in I$ let its unique sequential item-id be $sid(i) \in \{1, \ldots |I|\}$. For every user $u \in U$, we compute and store the (sparse) vector $h_{(u)}$ with dimensions equal to $|I|$, whose j-th component $(j = 1 \ldots |I|)$ is defined according to the equation

$$\left(h_{(u)}\right)_j = \sum_{p_{u,i} \in H_u : sid(i)=j} [m_{u,i} + 1] \tag{3.4}$$

Since relevance judgments are not available, we can use only price as the closest indication of user preference for certain items. Although the computational results showed that this formula improves quality of results, still the drawback of this approach is that prices are solely determined by the service provider, and do not reflect true relevance judgment of the specific user.

Using these vectors, we build for each item $i \in I$ another (sparse) vector $g_{(i)}$ with dimensions equal to $|U|$ whose j-th component $(j = 1 \ldots |U|)$ is defined to be

$$\left(g_{(i)}\right)_j = \begin{cases} \frac{1}{\sqrt{|H_u|}}, & sid(u) = j \wedge p_{u,i} \in H_u \\ 0, & \text{else} \end{cases} \tag{3.5}$$

where $|H_u|$ denotes the number of purchases user u has made so far.

Having these data structures available in shared memory, a number of threads are then spawned and execute in parallel without any further synchronization required, to compute for each item they have been assigned to, the k most similar items to it, together with their corresponding similarity values. Following loosely the SUGGEST recommendation library implementation (Karypis, 2001; Deshpande and Karypis, 2004), we define the similarity $sim\,(i_1, i_2)$ between two items i_1, i_2 to be the following quantity:

$$sim\,(i_1, i_2) = \frac{\sum_{j:(g_{(i_1)})_j > 0} \left(g_{(i_2)}\right)_j}{|g_{(i_1)}| \cdot \sqrt{|g_{(i_2)}|}} \tag{3.6}$$

where $|g|$ denotes the number of non-zero components of the vector g (notice how the similarity relationship between two items fails to be reflective, i.e., $sim\,(i_1, i_2) \neq sim\,(i_2, i_1)$ for $i_1 \neq i_2$ in general). This computation is fully parallelized in an *"embarrassingly parallel"* loop since no communication or synchronization between the threads is required.

Having computed (in parallel) and stored for each item, the k most similar items' indices, and their corresponding similarity values, the *k-NN-item-based recommender* computes the top-n recommendations for a user u, using the following procedure: for each non-zero element of the vector $h_{(u)}$, i, the k most similar items to i are examined, and those that are available and not already purchased by the user are added to a hash-table C_u whose keys are items j and values the sum of the quantities $(h_{(u)})_{sid(j)}/\sqrt{q}$ where q denotes the position in the list of k most similar items to j that item i is found in. Once all the non-zero elements of $h_{(u)}$ have been examined, the n key-value pairs in C_u with the highest values are proposed as the top-n recommendations for the user u.

3.2.4 User-Based Recommender

Our custom multi-threaded implementation of the *k-NN-user-based rec-ommender* is completely analogous to our custom implementation of the *k-NN-item-based recommender*. For every user $u \in U$, we define the (sparse) vector $\hat{h}_{(u)}$ in $|I|$ dimensions, whose j-th component ($j = 1 \ldots |I|$) is simply defined to be 1 if item i satisfying $sid(i) = j$ was purchased by the user, and 0 otherwise. Having obtained these vectors in a global shared memory, a number of threads are spawned that independently and concurrently execute in another embarrassingly parallel loop that does not require any synchro-nization or communication among them. The loop in each thread computes for each of a set of users it has been assigned to the similarity between this user and every other user in the database, according to the cosine-similarity formula $sim(u_1, u_2) = \hat{h}_{(u_1)} \cdot \hat{h}_{(u_2)}/(\|\hat{h}_{(u_1)}\|\|\hat{h}_{(u_2)}\|)$ (notice the reflective relationship that holds in this definition of similarity between users: the "amount of similarity" that u_1 has with u_2 is the same as that of u_2 to u_1). Once the $k(=150)$ most similar users to the given user u have been computed along with their similarity scores, these top k similarity scores are normalized to sum up to unity (by dividing each score by the sum of the k scores). These k most similar users to u define the *k-Nearest-Neighbors* of u.

Having created the above data structures in shared memory, our *k-NN-user-based recommender* computes the top-n recommendations for a given user u by computing for each (available and not already purchased) item in the history of the k most similar users to u the sum of the (normalized) similarity scores of the users that purchased that item; the algorithm then simply recommends the n highest scoring items to user u.

3.2.5 Final Hybrid Parallel Recommender Ensemble

The final top-n recommendations for a particular user u are computed by first asking each of the three recommenders (in parallel) to compute the top-$5n$ recommendations for u and then computing for each recommended item (by any of the individual recommenders) a linear weighted combination of the recommendation values of all three recommenders, whereby if an item is not in the top-$5n$ list of some recommender, it assumes by default the value zero for this recommender. The weights w^i, w^u, w^c of the item-based, user-based, and content-based recommenders were set (using the same optimization process that was employed for the computation of optimal weights for the parameters of the content-based recommender) to values approximately equal to 0.75, 0.15, and 0.1, respectively. The resulting values are sorted in descending order, and the top-n items are returned. The same linear-weighted combination process (with the same weights) applies when the recommender ensemble is asked to produce the final value of a (*user-id, item-id*) pair recommendation [see Amolochitis et al. (2013) for a detailed discussion of fusing ordered lists of search results of various heuristics in an ensemble to produce superior final ordered result lists].

3.2.6 Experiments with Other Base Recommender Algorithms

Two quite different base recommender algorithms are also very popular today. The first is the so-called *SlopeOne* recommender algorithm (Owen et al., 2011), which is not applicable in our case as it works only with datasets containing explicit user ratings of items. The second is reduced-dimensionality-based recommenders using *Singular Value Decomposition* (SVD, originally proposed as a method to make recommender systems more scalable in the face of very large datasets). Since our dataset is more than 99% sparse, we expected that SVD-based top-n recommendation results on this dataset would be inferior to the results of k-NN-based algorithms, as Sarwar et al. (2000) had reported previously. Indeed, the results produced by Apache Mahout's SVD Recommender implementation were quite worse than the results obtained by the other Boolean user-based recommender implementations available in Mahout, and for this reason, we do not investigate their use any further [similar quality results were produced using the Open-Source Funk SVD implementation (Ekstrand et al., 2011)].

3.3 Computational Results

In this section, we compare the performance of the introduced recommender ensemble with that of a Boolean recommender provided by the Apache Mahout Machine Learning platform. Due to the fact that no relevance judgments exist in our available data, we chose to use a Boolean recommender of Apache Mahout which requires no relevance judgments whatsoever, and specifically we have used a Boolean recommender algorithm with Log-Likelihood similarity measure and threshold-based definition of user-neighborhood, with threshold set at 0.3 (that we found to be the optimal threshold level for our data-set) as well as the individual performance of each of the three recommenders participating in our ensemble.

In order to evaluate the performance of the Mahout-based recommenders, we are have used the recall metric $R(10)$ as defined in Karypis (2001). Recall (together with so-called *precision-at-n* metric) is considered to be an appropriate metric in order to evaluate top-*n* recommendation results similar to our case.

The experiment was performed under the following configuration: for each specific user in the database, we have removed a single, *randomly chosen* item from the user's watching history and then apply the recommender ensemble in order to generate the top-10 recommendations for each user. In case that the top-10 recommendations include the removed item, then the objective function value is increased by one. The final objective function value, forming the $R(10)$ value, is the resulting sum divided by the total number of users in the database that had an item removed. With the given definition of recall, and test-bed construction, the average *precision-at-n* $P(n)$ satisfies $P(n) = R(n)/n$.

In order to optimize the objective function using a standard *alternating-variables* optimization process, we have used the popt4jlib Open Source library.

All test runs reported below were performed on a desktop PC with Intel Core-2 Quad CPU running at 2.4 GHz having 2-GB RAM running Windows. The testing dataset, being a snapshot of the database taken on April, 2013, comprises more than 20,000 users in total, with a little more than one million purchases (views) in total. The total number of items in the database is a little less than 7,000, but it is worth noting that the service provider's database contains a significant number of duplicate entries (entries with different item-ids for items with the same title, year of production, actors, directors, etc., with the possible exception that the genres in one entry are sometimes a subset of the genres in the other entry) that we had to keep track of, so that we never recommend an item that the user has already purchased, even though it is quite common in this dataset for the same user account to have purchased the same item many times (often 10 times or more); this holds especially true for items that belong to genres such as "Mickey Mouse' Fun Club" and others that are available free of charge. The user-item matrix's non-zero entries are less than 0.9% of the total number of cells in the matrix.

Table 3.1 provides the recall $R(n)$ values and associated running times T_n for the final ensemble, its individual recommenders acting alone, and Apache Mahout, for n = 10, 20, 30, … 100, for recommendations produced using the entire history of each user, except a single item randomly chosen from each user's history to act as the "hidden" item to measure recall against (Karypis, 2001).

A graphical illustration of the above results is shown in Figures 3.2 and 3.3, showing recall and response times of the various recommenders. Quite surprisingly, Apache Mahout's user-based recommender lacks very

Table 3.1 Comparing recommenders' quality and response times given the entire user histories (April, 2013)

	AMORE Ensemble			Apache Mahout			AMORE Item-based			AMORE User-based			AMORE content-based		
		Time (secs)			Time (secs)			Time (secs)			Time (secs)			Time (secs)	
n	R(n)	Load	Rec.	R(n)	Load	Rec.	R(n)	Load	Rec.	R(n)	Load	Rec.	R(n)	Load	Rec.
10	0,286	238	1076	0,158	238	17821	0,263	238	325	0,250	238	446	0,046	238	362,4
20	0,388	238	1176	0,232	238	17558	0,365	238	324	0,339	238	427	0,069	238	358,8
30	0,460	238	1325	0,286	238	17061	0,433	238	327	0,400	238	431	0,085	238	361,6
40	0,518	238	1452	0,326	238	17068	0,490	238	322	0,447	238	640	0,100	238	366
50	0,565	238	1618	0,367	238	17083	0,531	238	321	0,484	238	462	0,115	238	369,2
60	0,602	238	1729	0,403	238	17063	0,566	238	323	0,515	238	462	0,128	238	379,2
70	0,633	238	1752	0,435	238	17047	0,591	238	371	0,542	238	476	0,140	238	382,8
80	0,659	238	1989	0,465	238	17089	0,613	238	375	0,566	238	471	0,151	238	392,8
90	0,683	238	2161	0,489	238	17112	0,635	238	325	0,588	238	475	0,160	238	401,6
100	0,706	238	2814	0,512	238	17160	0,653	238	323	0,605	238	485	0,170	238	436,4

Recall Metric Comparison

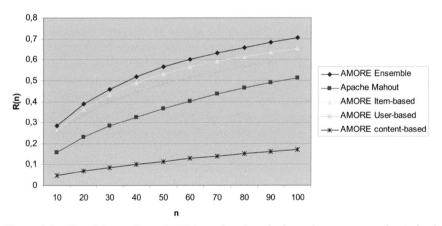

Figure 3.2 Plot of the recall metric $R(n)$ as a function of n for various recommenders trained on the entire user purchase histories.

Response Time Comparison

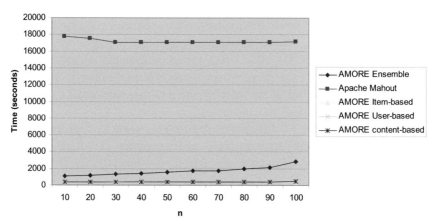

Figure 3.3   Plot of the response time as a function of *n* for various recommenders trained on the entire user purchase histories.

significantly behind both the AMORE ensemble, as well as our custom implementation of the user- and item-based recommenders in terms of recall (and equivalently, precision), as well as response times, and it is better than only our content-based recommender in terms of recall (but is much slower). This pattern holds for all values of *n*. As it can be easily verified, our AMORE ensemble is more than 80% better than Mahout in terms of the $R(10)$ metric, and is about 15 times faster than Mahout.

The ensemble's $R(10)$ value for those users whose history of purchases includes 50 or more items is 0.316, quite above the overall recall value of 0.28575, implying that for low values of *n,* the system is able to better understand the preferences of users with a large history of purchases. However, the ensemble $R(30)$ value is 0.475 for those users having made 50 purchases or more, which is now much closer to the overall $R(30)$ value of 0.45995, showing that as *n* gets larger, the recall value for the ensemble is approximately the same between users with small purchase histories and those with large ones.

Notice that the recall values obtained compare well with the best values obtained for much more controlled datasets, such as the *movie-lens* dataset where the ratings' information that is made available for each user is the true rating the particular user has given to the item, as opposed to our dataset that contains only the purchase history of each user account (that is often

Table 3.2 Comparing recommenders' quality and response times given the history of user purchases that occurred between 9pm and 1am (dataset of April, 2013)

	AMORE Ensemble	Time (secs)		Apache Mahout	Time (secs)		AMORE Item-based	Time (secs)		AMORE User-based	Time (secs)		AMORE content-based	Time (secs)	
n	R(n)	Load	Rec.	R(n)	Load	Rec.	R(n)	Load	Rec.	R(n)	Load	Rec.	R(n)	Load	Rec.
10	0,243	241	498	0,165	241	4531	0,223	241	243	0,202	241	259	0,062	241	999
20	0,334	241	578	0,237	241	4529	0,314	241	242	0,282	241	401	0,089	241	1009
30	0,401	241	704	0,293	241	4431	0,379	241	253	0,335	241	399	0,111	241	1026
40	0,453	241	872	0,339	241	4482	0,428	241	243	0,381	241	396	0,151	241	1008
50	0,495	241	1043	0,373	241	4459	0,466	241	246	0,416	241	399	0,151	241	1061
60	0,534	241	1040	0,409	241	4489	0,496	241	240	0,444	241	401	0,167	241	1042
70	0,566	241	1272	0,439	241	4434	0,521	241	244	0,471	241	404	0,183	241	1054
80	0,593	241	1507	0,470	241	4435	0,543	241	247	0,495	241	403	0,195	241	1076
90	0,620	241	1798	0,499	241	4444	0,562	241	246	0,514	241	405	0,208	241	1094
100	0,642	241	2170	0,518	241	4524	0,578	241	247	0,533	241	405	0,217	241	1128

used by all members of the household). To alleviate this additional problem with our dataset, we have provided an additional feature to our algorithms, namely, the ability to train them using only those content-items seen by the user within a particular time-window. The rationale behind this choice is that by narrowing the user history to items seen, for example, during prime-time, the chances that this user history is the union of more than one actual person in the household should be reduced, and therefore, the accuracy of the system should be increased. In Table 3.2, we show the results of running the various recommenders trained using only items that were seen by the users during a time that overlaps with the "prime-time" window between 9pm and 1am. The results again show a very clear superiority of our ensemble, even though they do not improve upon the results obtained when training the classifiers with the entire history of user purchases, thereby the hypothesis that time-windows can help narrow down the persons using the service from each user account does not have statistical support. The quality of the results is visualized in Figure 3.4. Regarding running times, our ensemble is between 1.98 and 6.46 times faster than Apache Mahout.

We attribute the much faster response times of our system to two main reasons:

1. *A sophisticated multi-threading design and implementation* that allows the software to utilize 100% of the available cores of the CPU and obtain essentially linear speedups. To achieve this performance, each running thread never creates any objects on the heap (that dramatically reduce parallel performance) using the operator *new* and, of course, does not have to obtain any synchronization locks as they only write data in different areas of the same arrays and do *not* require any data computed in parallel from the other threads.

Recall Metric Comparison

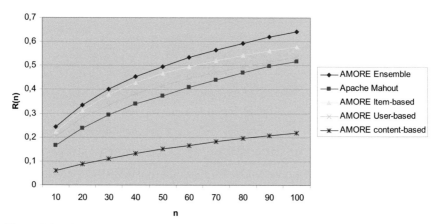

Figure 3.4   Plots of the recall metric *R(n)* as a function of *n* for various recommenders trained on user histories on the interval 9pm to 1am.

2. *A better-suited implementation of sparse vectors* for *k*-NN-based implementations of recommender algorithms than the one available in the Colt numeric library that was adopted for Apache Mahout's core numeric computations, combined with a very fast implementation of thread-local object pools for light-weight objects that make it possible for the computing threads never to call the *new* operator as stated in reason #1 above.

As another experiment, we have deleted from the snapshot of our database taken on April, 2013, all user purchases that occurred during the last two weeks recorded in the system, and have trained the system with the remaining older data, to see the levels of the precision and recall metrics on this differently constructed test dataset. The plots in Figure 3.5 show how average *precision, recall,* and the combined *F-metric* vary with different recommendation list lengths (measured in points that are multiples of 5 and 8). The reduced recall values are expected since the system must now be able to find not just one of the items the user has selected at any random point in the past, but the items the user saw in the last two weeks: but within the last 2 weeks, items made available within that time-frame may have not been seen yet by a statistically significant number of users so that the system can "understand" to what other items they are similar with, thus the drop in the recall values.

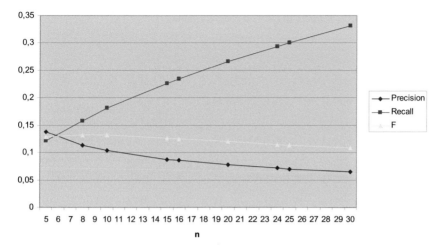

Figure 3.5 Plots of *precision*, *recall*, and *F-metric* for the AMORE ensemble when the test data are the last 2 weeks of user purchases. The *F-metric* is maximized at $n = 10$.

We have performed an *empirical* small-scale test where we asked eight volunteer users (other than the authors) to explicitly state the relevance (like/dislike) of the top-10 recommendations the systems produced for them, *after declaring just five of their favorite movies*. The precision of the results is shown in Figure 3.6, and is much more encouraging. The significant

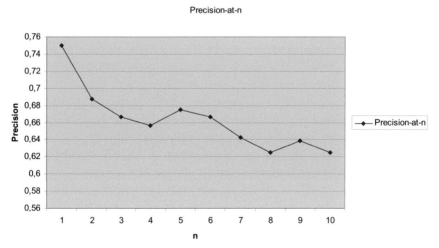

Figure 3.6 Empirical average AMORE *precision-at-n* measured after users have stated exactly 5 of their most favorite movies.

Recall Curves Temporal Evolution

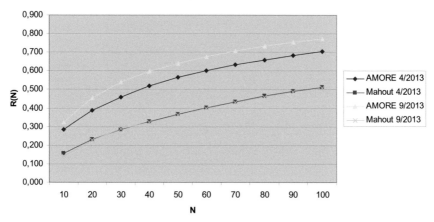

Figure 3.7   Temporal evolution of AMORE and Mahout performance.

difference between *explicitly stated user-relevance* and *calculated system accuracy from user histories* can be attributed to many factors, the most prominent of which would be the fact that users are very likely to have already seen in the theaters their favorite movies that the system calculates for them, or the sometimes high pricing of specific content items available for viewing.

Finally, in Figure 3.7, we show how AMORE performance has evolved over time.

The latest experimental results on system recall and response times (September, 2013, on a database of more than 26,000 users and more than 1.9 million views) show that AMORE outperforms Apache Mahout by more than 100% in terms of the $R(10)$ metric and more than 6,300% in terms of speed! AMORE has been increasing its performance as time passes by, by more than 13.8% between April and September, 2013. Mahout's user-based recommender (using the Log-Likelihood metric) on the other hand dropped its performance by more than 10% in the same time interval.

Cremonesi and Turrin (2009) and Bambini et al. (2010) showed that in their own production environments, the recall rate of item-based recommenders may deteriorate as time passes by, due to cold start issues and the fact that once new users view so-called "easy-to-recommend" items (i.e., blockbusters), the task of the recommender engine becomes much more difficult. In contrast, our results indicate that the combination of our custom item-based recommender, user-based, and content-based leads to a system that evolves

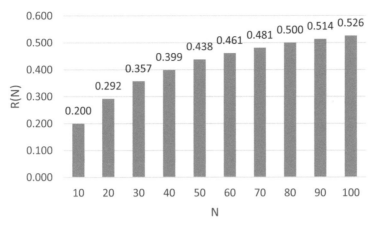

Figure 3.8 Recall metric *R(n)* using an alternative ensemble (consisted of: (i) content- and (ii) item-based recommenders).

so that it improves its recall rate as time passes, and the improvement is significant.

Finally, we have performed a series of experiments using an alternative ensemble configuration consisted of the: (i) *content-* and (ii) *item-*based recommender (instead of the three recommenders of the original ensemble). We have measured the performance of the particular ensemble using the *R(n)* metric and we depict the results in Figure 3.8.

In Table 3.3, we compare the performance of the original (*content, item,* and *user* based) ensemble, with the alternative (*content* and *item* based).

Table 3.3 Comparing the original (*content, item, and user*) AMORE ensemble with the alternative (*content and item* based)

	R(n)	
n	AMORE Ensemble	Content, Item Based Ensemble
10	0.286	0.200
20	0.388	0.292
30	0.460	0.357
40	0.518	0.399
50	0.565	0.438
60	0.602	0.461
70	0.633	0.481
80	0.659	0.500
90	0.683	0.514
100	0.706	0.526

The results show that the original 3-recommender ensemble outperforms the alternative 2-recommender in terms of the $R(n)$ metric for all different values of n.

3.4 User and System Interfaces

AMORE recommendations are shown to the service subscribers on their TV screen in a special screen shown in Figure 3.9. The first row shows the recommended movies for the user account, whereas the second row immediately below it shows the month's most popular movies.

While Figure 3.8 shows a screen-shot of the user interface as seen by the end-user (the service subscriber), AMORE offers a variety of other interfaces. In Figure 3.10, we show a SOAP-UI snapshot of the WSDL interface that AMORE exposes to its consumers that simply consists of item recommendations for a particular user subject to certain constraints (such as time-window constraints, item-availability constraints, etc.).

Figure 3.9   AMORE end-user on TV–screen interface.

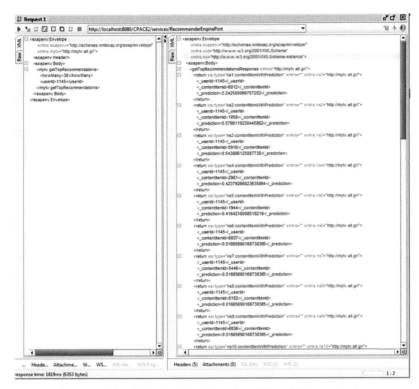

Figure 3.10 AMORE WSDL interface (SOAP-UI screenshot).

Figure 3.11 AMORE developer desktop UI.

In Figure 3.11, we show a developer GUI, developed particularly for the purposes of easier testing of the recommendations produced by AMORE. The developer GUI connects to the same databases that AMORE connects and consumes the AMORE web services, but allows the user to see much more quickly most information relevant to a user and their past choices as well as the recommendations the system makes for them, thus allowing easier testing and validation of the system. This GUI was the tool used to measure the empirical performance of the system regarding *precision-at-n P(n)* for our eight volunteer users.

4

Quantitative Association Rules Mining

4.1 Why Quantitative Association Rules?

Quantitative association rules (QARs) refer to a special type of association rule of the form of *S* implies *I*, where *S* is the rule's *antecedent* and *I* is the rule's consequent, both of which consist of a set of *numerical* or *quantitative* attributes. In contrast, *Boolean* association rules, also of the form *S* implies *I*, are consisted of categorical (nominal or discrete) attributes. The *quantitative* aspect of the association rules allows for a wide range of application in different types of settings containing transactional data. Such transactional data can be found, for instance, in *e-commerce* applications.

During recent years, there has been an emergence of such e-commerce services (Kotsiantis et al., 2006), which signals a need for efficient recommendation algorithms that address the various issues that are raised. In such a setting, traditional data mining techniques, including association rules mining, are very frequently used (Sarwar et al., 2000) and as the amount of available, consumable content increases, *price* proves to be, in many cases, a determining factor that motivates certain consuming behavior.

Furthermore, certain *online* services, including video-on-demand services offered by triple play providers, which charge a certain monthly fee to users of their services have an even more limited potential of making significant additional profits (beyond income coming from subscription fees) since users usually expect to get most movies for *free*, since they have already paid in advance for the benefits of the service. Therefore, in such a setting, which contains users who prove to be skeptical in paying additional charges for streaming a movie, price is a very important factor that would determine their decision to consume the item, or not.

Being able to model *patterns* of consuming behavior, in relation to certain prices (deemed to be *attractive* by users), provides valuable knowledge for the design of efficient recommender systems. Item associations that take into consideration numerical attributes, like price, can be efficiently represented via QARs, and mining such rules proves to be a very powerful recommendation technique.

We have introduced an algorithm for mining QARs by processing a number of user histories from transactional databases containing numeric attributes, in order to generate a set of association rules with a minimum support and confidence value. The generated rules show strong relationships between certain items that have been consumed at specific price levels, information that is used in a novel recommendation post-processing algorithm, that uses the generated association rules in order to improve the quality (in terms of recall) of an original set of recommendations. We have experimented extensively with available production data from a major triple play services provider, as well as a publically available dataset.

Finally, we have implemented a custom synthetic dataset generator (SDG) that allows for the generation of different datasets (under different parameters), simulating e-commerce related scenarios with respect the number of users and their respective number of transactions in different *cycles,* as well as price fluctuations based on demand changes.

4.2 Algorithm Overview

In this section, the design of the Quantitative Association Rules Mining (QARM) algorithm is presented.

Specifically, QARM is an algorithm for computing all Quantitative Association Rules (QARs) from a transactional database D containing "user histories" of the form: user u purchased a set of $\{i, p\}$ pairs, where i corresponds to the item purchased and p corresponds to the price at which item i has been purchased. All generated QARs are of the form: "IF a user has paid at least $p_{A_1} \ldots p_{A_k}$ for products $A_1 \ldots A_k$, THEN such a user would be willing to pay p_C for product C.

Apart from the database of "user histories," the algorithms requires as input a minimum required *support* value s as well as a minimum required *confidence* value c. The output of the algorithm includes all QARs of the aforementioned form having support and confidence above the minimum specified thresholds.

4.3 Algorithm Design

A detailed description of the QARM algorithm follows.

First, QARM calls the method computePriceLevelsIncreasingOrder(D) that computes all different price levels p that occur in D and sorts them in increasing order $0 \leq p_1 < p_2 \cdots < p_n$. The total number of distinct price levels n, which is equal to the *size of* set p, is returned by the method sizeof (p).

Next, the algorithm uses the FP-Growth algorithm to generate all frequent item-sets F_s with support s or higher, assuming zero price level, i.e., by treating the database as a qualitative database where all prices paid were exactly 0.

The algorithm then defines the following sets: R containing all QAR that are *mined* by the algorithm, and set C containing all candidate association rules that the algorithm needs to examine. Specifically, set C is populated as follows. For each i, in each frequent item set f in the set of F_s a new candidate rules is formed. So for each item i in the frequent item set f, the algorithms forms a new candidate rule where i is placed as the consequent I of the rule, and all other items in f are placed as antecedents S of the rule. The newly formed candidate rule is then placed in set C. Note that the algorithm considers only those frequent item sets f where the total number of items is higher than 1.

After set C is populated, QARM performs for each candidate rule r that is member of C the following process.

Defines set T which is a stack of sets of pairs of the form $(A \in \ell_k, p \in \{p_1 \ldots p_n\})$.

For each price level p_i of p (in *decreasing* order), QARM sets the price of consequent item I to p_i by defining a set Q consisted of pairs of items of the form (I, p_i).

Definition 3.1: *SUPPORT* ($r \, | Q$)

SUPPORT ($r \, | Q$) calculates the *support* value s for a rule ($r = S \rightarrow I | Q$) equals the number of users that purchased each of the items in $S \cup \{I\}$ at a price at least equal to the price specified for that item in Q *divided* by the total number of user histories in the database. If for an item, no price is specified in Q, the condition is that the user has purchased the item (for free or for any price).

Quantitative Association Rules Mining invokes method *SUPPORT* ($r \, | Q$) and if the constraint $SUPPORT(r|Q) < s$ holds, meaning that the output of the *SUPPORT* method is below the minimum required support threshold s, then the algorithm *skips* the current iteration and proceeds to the next value for p_i.

Alternatively, the algorithm continues with the current iteration and pushes Q onto T. While T is not empty, the algorithm pops a Q from T and iterates over each element J contained in set S that is not already part of set Q, and for each such element, it further iterates over all different price levels values (sorted in *ascending* order). The algorithm then creates set Q' which is consisted of all elements in set Q in addition to a new pair of item J at a price specified in the current iteration step (see previous statement).

Definition 3.2: *CONFIDENCE*: A rule $(r = S \rightarrow I|Q)$

A rule $(r = S \rightarrow I|Q)$ has *confidence c* that equals the number of users that purchased each of the items in $S \cup \{I\}$ at a price at least equal to the price specified for that item in Q *divided* by the number of users that purchased each of the items in S at a price at least equal to the price specified for that item in Q. If for an item, no price is specified in Q, the condition is that the user has purchased the item (for free or for any price).

Returning back to the description of QARM, and the main algorithm flow, if the constraint $SUPPORT(r|Q') \geq s \wedge CONFIDENCE(r|Q') \geq c(1)$ holds, meaning that the output of the *SUPPORT* method is above the minimum required support threshold *s as well as* the output of the *CONFIDENCE* method is above the minimum required confidence threshold *c*, then the algorithm *proceeds*; otherwise, the algorithm examines whether the condition $SUPPORT(r|Q') < s$ holds and if so, *breaks* the current iteration, returning to the iterator that examines next J item, or otherwise, the algorithm control returns to the *ascending price* iterator.

Returning back to the normal flow where condition (*1*) holds, the algorithm then calls method Add-non-dominated $(r|Q')$ to R and pushes Q' onto set T.

Definition 3.3: *Add-non-dominated $(r|Q)$ to R*

The operation *"Add-non-dominated $(r|Q)$ to R"* will add the QAR $(r|Q)$ into the QAR-set R only if R does not contain already another rule $(r'|Q')$ that dominates the rule to be added. A rule $(r = S \rightarrow I|Q)$ is dominated by another rule $(r' = S' \rightarrow I, Q')$ if:

$S \supseteq S'$ and the consequent's price for r' is higher than the consequent's price in $(r|Q)$ and

$$SUPPORT(r|Q) \leq SUPPORT(r'|Q') \text{ and}$$

$CONFIDENCE(r|Q) \leq CONFIDENCE(r'|Q')$ and

$$\forall (A, p') \in S' : (A, p) \in S \rightarrow p' \leq p.$$

When a rule $(r|Q)$ is added onto R, the function must also ensure that it will remove from the set R any rules dominated by $(r|Q)$. The algorithm then proceeds with all subsequent iterations up until all candidate rules are processed, or a specified *maximum* limit of number of rules to process is reached.

A *code* representation of the QARM algorithm follows.

Definition 3.4: *QARM algorithm*

Begin

0. Let p all different price levels that occur in D in ascending order $0 \leq p_1 < p_2 \ldots < p_n$.
1. Let F_s the set of all frequent item-sets F_s
2. Let $C = \emptyset; R = \emptyset$.
3. foreach frequent k-itemset $\ell_k \in F_s, k \geq 2$ do:
   3.1. Create the set $H_1 = \{r = (S \rightarrow I) : S = \ell_k - I, I \in \ell_k\}$.
   3.2. Set $C = C \cup H_1$.
4. endfor.
5. foreach rule $r = (S \rightarrow I) \in C$ do:
   5.1. Let $T = \emptyset$. // T is a stack of sets of pairs of the form $(A \in \ell_k, p \in \{p_1 \ldots p_n\})$
   5.2. foreach $i = n \ldots 1$ do:
   5.2.1. Let $Q = \{(I, p_i)\}$. // set the price of I to p_i.
   5.2.2. if $SUPPORT(r|Q) < s$ continue;
   5.2.3. Push Q onto T.
   5.2.4. while $T \neq \emptyset$ do:
       5.2.4.1. Pop a Q from T.
       5.2.4.2. foreach $J \in S - \underset{(A,p) \in Q}{\cup} A$ do:
           5.2.4.2.1. foreach $j = 1 \ldots n$ do:
               5.2.4.2.1.1. Let $Q' = Q \cup \{(J, p_j)\}$.
               5.2.4.2.1.2. if $SUPPORT(r|Q') \geq s \wedge$
                           $CONFIDENCE(r|Q') \geq c$
               5.2.4.2.1.2.1. Add-non-dominated $(r|Q')$ to R.
               5.2.4.2.1.2.1. Push Q' onto T.
               5.2.4.2.1.3. elseif $SUPPORT(r|Q') < s$ break.

5.2.4.2.1.4. endif.

5.2.4.2.2. endfor. // j

5.2.4.3. endfor. // J

5.2.5. endwhile. // T

5.3. endfor. // i

6. endfor. // r

7. return R.

4.4 Recommender Post-Processor

4.4.1 Overview

The original set of recommendation results can be enhanced with the use of a Post-Processor that updates the results based on knowledge extracted from the generated set of QARs under minimum *support* and *confidence* values. The Post-Processor aims to promote certain recommendations that are part of the consequent of QARs that *fire* for a specific user and, therefore, are considered to be *relevant* for the specific user. The number of positions that a consequent item is promoted in the original recommendation list depends on the *confidence* value c of the rule as well as whether the item exists in the original recommendation list. Specifically, an item that is recommended by both a fired rule and a recommender is promoted at higher positions (due to the increased confidence that the item is indeed relevant to a user) than an item that is only part of a fired rule. Also the confidence value of the fired rule boosts the number of positions to promote. Nevertheless, the algorithm considers that items contained in fired rules, to be of high relevance, therefore promote them whatsoever in the expense of other items contained at the lowest ranks of the original recommendation list.

4.4.2 Post-Processing Algorithm

Given a set of recommendations L generated for user u and a set of QARs R, a first, the algorithm defines as *orgIdx* the index of the consequent item in the original recommendation list. In case the item is not contained in the original list, the value of *orgIdx* is equal to "−1"; otherwise, *orgIdx* has a value greater or equal to zero, and less than or equal to the length of the original recommendation list minus "1" (assuming a 0-*based* indexing scheme). Then the value of *newIdx* is calculated by calling calcNewIdx which receives as parameters *orgIdx*, the rule's confidence value c, as well as the size of the original recommendation list, *recListSize*. The value of *newIdx* corresponds to the index position that the consequent item is going to be placed in the original list, and is calculated as follows:

Define method calcNewIdx (*orgIdx, c, recListSize*).

  0. Let *isNew* ← *false*.
  1. if *orgIdx* = −1 then

   a. *isNew* ← *true*.

  2. end if.
  3. Let *addPosToPromote* ← *isNew* ← 7 : 8.
  4. Let *totalPosToPromote* ← calcNumOfPosToPromote(*c*) + *addPosTo-Promote*.
  5. Let *startIdx* ← *isNew* ← *recListSize* −1 : *orgIdx*.
  6. Let *newIdx* ← *startIdx* − *totalPosToPromote*.
  7. if *newIdx* < 0 then

   a. *newIdx* ← 0.

  8. end if
  9. return *newIdx*.

Define method calcNumOfPosToPromote (*c*).

  1. Let *confMin* ← 0.4.
  2. Let *confThreshold* ← 1.5.
  3. Let *confThresholdStep* ← 0.2.
  4. Let *confRatio* ← *c*/*confMin*.
  5. Let *confRatioDelta* ← *confRatio* − *confThreshold*
  6. Let *numOfPos* ← *confRatioDelta*/*confThresholdStep*.
  7. return *numOfPos*.

After the Post-Processor calculates the values of *orgIdx* and *newIdx,* respectively, it applies the re-ordering of the original recommendation list by *moving* the item, originally at position *orgIdx* to the position *newIdx*, pushing at the same time all elements with original index less or equal to *newIdx* one index position lower, eventually causing the *last* item (in the original list) to be eliminated.

4.5 Synthetic Dataset Generator

The proposed QARM algorithm has high applicability in different *e-commerce* settings, and therefore we performed a series of simulations for different e-commerce-related scenarios. These simulations allowed the evaluation of the performance of QARM under different datasets with specific characteristics as we will explain in the following section. We have designed and implemented an *SDG*, a program that allows the generation of datasets according to specific predefined configuration parameters.

4.6 Configuration Parameters

The SDG configuration parameters, which are specified prior to the generation of a dataset instance, include information such as: (i) the number of *users* as well as (ii) the number of *items* that exist in the dataset, (iii) the number of *cycles* to run, i.e., arbitrarily defined, short-term periods during which items are consumed (say, weekly or biweekly), and (iv) the *maximum number of items* that may consumed by a user during a cycle. Additional configuration parameters include: (v) the different *price levels*, i.e., an ordered collection of distinct price values that are randomly assigned to each generated item, and (vi) the different maximum reference price levels, i.e., an ordered collection of distinct maximum price values that are randomly assigned to each generated user, representing the maximum price that the specific user is willing to pay for every available elastic item.

4.7 Item Demand Elasticity

All generated items are distinguished into one of two main categories: (a) *elastic* and (b) *inelastic* items.

We consider as *inelastic*, items that are deemed of *high popularity*, to the point that price has absolutely no effect on their high demand levels. So, *inelastic* items will be consumed by a user regardless of the maximum reference price level constraint that a specific user has toward *elastic* items.

Contrary, we consider as *elastic*, items whose demand has a direct relationship to their price, and therefore, the criterion based on which these items are consumed (or not) is whether the item price is below (or equal to) the maximum reference price that a specific user is willing to pay for an *elastic* item.

4.8 Dataset Generation Process

Initially, SDG generates the configured number n of items I. All generated items have *id* values of integer type which is assigned using an incremental approach.

4. Set $I \leftarrow \{\}$.
5. Set $n \leftarrow 1000$.
6. Set $idx \leftarrow 1$.
7. foreach idx in n do

      a. Set $c \leftarrow$ createItem(idx).
      b. add (c) in I.

8. endfor
9. end.

Where createItem(idx) creates a new item instance with id value equal to the specified idx parameter.

Each generated item is assigned: (i) elasticity type (*inelastic* or *elastic*) and (ii) price. With respect to the elasticity type, our implementation offers two alternative ways of assigning values to items: either (a) creates a specific number of inelastic items that equals a configured percentage of the total set of generated items or (b) for each generated item, assigns the elasticity type by generating a random Boolean value. Finally, the price is randomly assigned from a specific fixed, ordered collection of distinct price values.

At a second step, SDG generates the configured number of users. We have introduced a structure named *user groups* which corresponds to the different user categories, each of which is characterized by a distinct maximum reference price value that each user belonging to the group is willing to pay for every available elastic item in the dataset. We then *randomly* place each of the generated users in one of the previously mentioned user groups so that they subscribe to the maximum reference price value of the group she belongs to.

At a third step, SDG generates a specific number of mock association rules, which are fully configurable with respect to the number of items that will form the antecedent of the rule (the consequent always has *one* item). Each mock rule is generated by randomly picking items to fill the antecedent and consequent of the rule, respectively. The only constraint is that all randomly picked items are not already part of the rule as part of either the antecedent or the consequent. Generated mock rules are then persisted for future reference.

At a fourth step, SDG runs the dataset population process using the generated *item* and *user* sets by simulating a specified number of transactions for each generated user for a specified number of cycles. Each cycle simulates an arbitrary short-term period during which a maximum, pre-configured number of items may be consumed. SDG runs the following simulation.

4.8.1 Generation Cycle

In the case of the first cycle, the program generates the first instance of the dataset. For each different user group, the program processes all users belonging to the group level. Each user in the group *randomly picks* a specified number of items from the collection of available items with the following logic.

Our simulation aims to model the real-case scenario where certain items are considered to be more attractive to users compared to other items. We model this variance in popularity with set *M*, which corresponds to a set of *weight* values for each respective item in set *I* (an element of set *M* is mapped based on its index with the respective item in set *I*). The *weight* value increases, as the *id* value increases, and therefore items with higher *id* values are considered as *more* popular than those with lower *id* values.

1. Set $M \leftarrow \{\}$.
2. Set $n \leftarrow 1000$.
3. Set $exp \leftarrow 0.6$.
4. Set $idx \leftarrow 1$.
5. foreach *idx* in *n* do

    a. Set $id \leftarrow$ getItemId(*I*, *idx*).
    b. Set $m \leftarrow id^{exp}$.
    c. add *m* to *M*.

6. endfor
7. end.

Where getItemId(*I*, *idx*) returns the id of the item at index *idx* in the set of items *I*.
Define method chooseRandomItemFromSet.

1. Set *r*.
2. Set $b \leftarrow -\infty$.
3. Set $d \leftarrow 20.0$.
4. Set $seed \leftarrow 7$.
5. Set $idx \leftarrow 1$.
6. foreach *m* in *M* do

    a. Set $v \leftarrow m +$ nextGaussian(*seed*) $* m / d$
    b. if $(v > b)$ then

        i. $r =$ getItemId(*I*, *idx*)
        ii. $b = v$.

    c. endif

    d. *idx* ← *idx* +1.

7. endfor

8. return *r*.

9. end.

Where *r* is the randomly picked item from the set of items *I* and nextGaussian(*seed*) returns the next pseudorandom, Gaussian ("normally") distributed value with mean 0.0 and standard deviation 1.0.

    In case that the user has picked an *inelastic* item, then the item is immediately consumed by the user. Contrary, in case that the user has picked an *elastic* item, then the program examines whether *the maximum reference price* specified in the group (that the specific user belongs to) is higher than or at least equal to the actual price of the picked item, and if so, the user consumes the item; otherwise, the item is discarded. When the aforementioned process completes for the first cycle, then an initial instance of the dataset is generated containing the user transaction history generated during the cycle.

4.8.2 Update Cycle

For all following cycles, SDG applies the update process logic, maintaining at the same time *snapshots* of the dataset for each specific cycle, and is then used for later reference. The update process is very similar to the generation process, in the sense that the steps followed are identical with the exception of one additional step applied in the beginning of the cycle (prior to performing the process for each user) that aims to adjust the item prices based on the items' change in demand. This last step is applied *only* while processing a cycle with index value greater than "2" (assuming a *1-based* indexing scheme).

    In the price adjustment step, SDG examines for each item, the demand change, *d*, as witnessed during the previous two cycles. So for each item, the program calculates *d*, which is defined to be the ratio of the absolute demand change over total user histories. At configuration parameter level, the lower and upper ratio limit values for *d* are specified. If the demand change, *d*, is below the lower limit, the item price is *lowered* by one unit, and similarly, if *d* is higher than the specified upper limit, its price is *increased* by one unit.

This latter step allows SDG to simulate some real-world consuming behavior where prices fluctuate as a function of their change in demand, and eventually reach a state of *equilibrium* (with respect to user's maximum reference price levels) after a sufficient number of cycles.

Finally, SDG iterates the structure containing all cycle-specific data to save the *entire* generated dataset into file.

4.9 Experimental Results

4.9.1 Metric

In order to evaluate the performance of the QARs mining algorithm over the synthetically generated datasets (under different parameter values), we use the *precision* metric, which in this context is defined as the ratio of the sum of all (rule) *hits* over the sum of all (rule) *hits* plus the sum of all (rule) *misses*:

$$Precision = \frac{\sum hits}{\sum hits + \sum misses}$$

We define that: a rule *r fires* for user *u* when all antecedent items *S* of *r* have been consumed by *u* at prices that are at least equal to the ones specified in *r*. For all rules *R* that fire for user *u*, we define as a *hit* those rules whose consequent set *I* (in our case, all consequents sets *I* are of size "1") contains an item that has been consumed by *u* at a price *at least equal to* the one specified in the rule. Similarly, we define as a *miss* those rules whose consequent contains an item that has been consumed by *u* at a price *above* the one specified in the rule. The sum values of total hits and total misses are calculated by examining the total number of hits and misses, respectively, for all active users in the database which are included in the current experiment.

4.9.2 QARM Results Using Synthetically Generated Datasets

In this section, we will present the experimental results for the QARM algorithm on a number of different synthetically generated datasets using different configuration parameters, which affected the dataset size and complexity. The experiments we have conducted show the performance of QARM both with respect to precision accuracy as well as processing time.

Table 4.1 Main dataset configuration parameters

Configuration	Number of Items	Number of Users	Number of Cycles	Purchases Per User Per Cycle	% of Elastic Items
1	2,000	2,000	10	10	51%
2	3,000	3,000	10	10	50%
3	1,000	1,000	100	1	52%

We have constructed different synthetically generated datasets using different configuration parameters. The main three types of configuration parameters are presented in Table 4.1. The parameter variables represent the following:

1. *Number of Items* represents the total number of available content items that can be consumed by users.
2. As previously mentioned, each item can be either *elastic* or *inelastic* and item prices are randomly chosen from a predefined set of 10 distinct price levels. *% of Elastic Items* specifies the percentage of the total set of items that is of type *elastic*.
3. *Number of Users* represents the total number of users in the dataset. Each user has a maximum reference price level which is assigned randomly from a predefined set of 10 distinct maximum price levels.
4. *Number of Cycles* represents the total number of cycles for which the dataset generation process will run. Cycles simulate small periods of time during which items are consumed. Within each cycle, each user can consume up to a specified maximum number of items, which is defined by the parameter *Purchases per User per Cycle*. The user *randomly* picks an aforementioned number of items, and for each picked item, the code examines whether the item is inelastic (and if so, it is consumed immediately despite the item's price); otherwise, the code examines whether the item's price is below or equal to the maximum reference price set for the specific user.

In Figures 4.1 to 4.4, we depict the performance of QARM under a synthetically generated dataset with the parameters specified in *Configuration* "1" in terms of precision accuracy and computational performance.

Figure 4.1 depicts the performance of QARM in terms of precision accuracy for a different set of minimum confidence values for a fixed minimum support value of *0.3*.

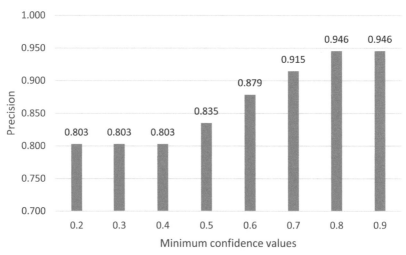

Figure 4.1   Precision of QARM with fixed support value under Configuration "1."

Figure 4.2 depicts the computational performance of QARM, in terms of minutes lapsed, for the experiment depicted in Figure 4.1.

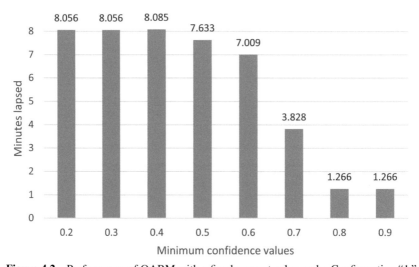

Figure 4.2   Performance of QARM with a fixed support value under Configuration "1."

Figure 4.3 depicts the performance of QARM in terms of precision accuracy for a different set of minimum support values for a fixed minimum confidence value of *0.6*.

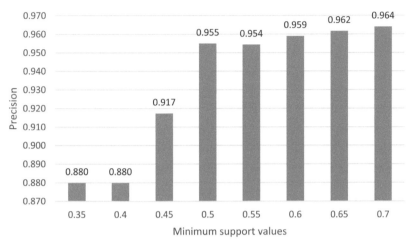

Figure 4.3 Precision of QARM with a fixed confidence value under Configuration "1."

Figure 4.4 depicts the computational performance of QARM, in terms of minutes lapsed, for the experiment depicted in Figure 4.3.

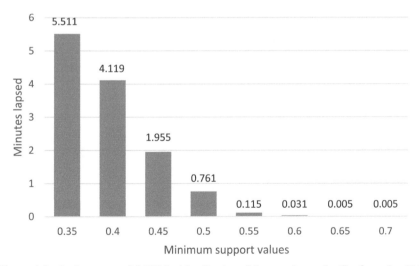

Figure 4.4 Performance of QARM with a fixed confidence value under Configuration "1."

In Figures 4.5 to 4.8, we depict the performance of QARM under a synthetically generated dataset with the parameters specified in *Configuration* "2" in terms of precision accuracy and computational performance.

Figure 4.5 depicts the performance of QARM in terms of precision accuracy for a different set of minimum confidence values for a fixed minimum support value of *0.3*.

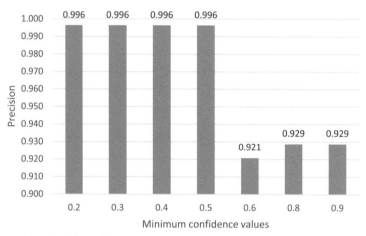

Figure 4.5   Precision of QARM with a fixed support value under Configuration "2."

Figure 4.6 depicts the computational performance of QARM, in terms of minutes lapsed, for the experiment depicted in Figure 4.5.

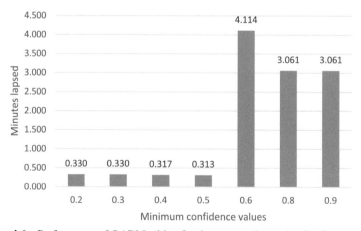

Figure 4.6   Performance of QARM with a fixed support value under Configuration "2."

Figure 4.7 depicts the performance of QARM in terms of precision accuracy for a different set of minimum support values for a fixed minimum confidence value of *0.6*.

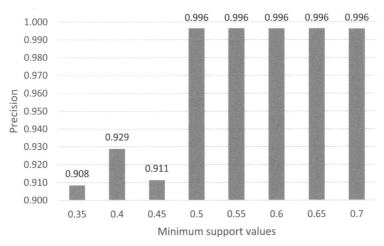

Figure 4.7   Precision of QARM with a fixed confidence value under Configuration "2."

Figure 4.8 depicts the computational performance of QARM, in terms of seconds lapsed, for the experiment depicted in Figure 4.7.

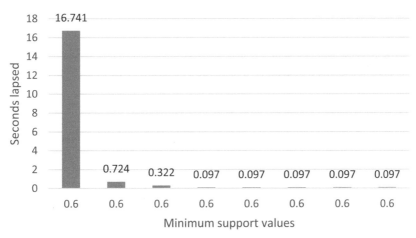

Figure 4.8   Performance of QARM with a fixed confidence value under Configuration "2."

In Figures 4.9 and 4.10, we depict the performance of QARM under a synthetically generated dataset with the parameters specified in *Configuration* "3" in terms of precision accuracy and computational performance.

Figure 4.9 depicts the performance of QARM in terms of precision accuracy for a different set of minimum confidence values for a fixed minimum support value of *0.7.*

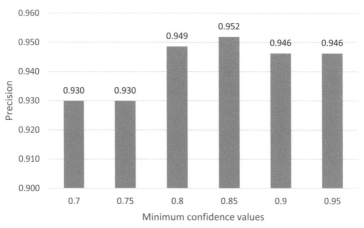

Figure 4.9   Precision of QARM with a fixed support value under Configuration "3."

Figure 4.10 depicts the computational performance of QARM, in terms of minutes lapsed, for the experiment depicted in Figure 4.9.

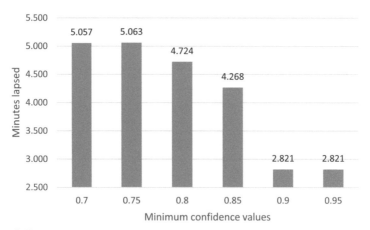

Figure 4.10   Performance of QARM with a fixed support value under Configuration "3."

4.9.3 QARM Results Using Movielens Dataset

We have also conducted a series of experiments for the QARM algorithm using the MovieLens dataset (specifically the version of the dataset containing 100,000 transactions) with different confidence values for a set of fixed support value. In this section, we will present the experimental results for the QARM algorithm on the aforementioned dataset with respect to precision accuracy. We will also present charts depicting the total candidate rules processed for each different case.

In Figures 4.11 to 4.18, we depict the variation in precision accuracy using different fixed minimum required support values ranging from *0.3* to *0.45* (with a "step" incremental value of *0.5*) and variable confidence values. Additionally, we depict the number of QARs generated under each such configuration.

4.9.4 QARM Results Using Post-Processor

We have performed a series of experiments to evaluate the percentage of improvement by using the Post-Processor on top of the results generated by the Hybrid (*item* and *content* based) recommender. We have conducted the experiment on production data provided by the Triple Play service provider.

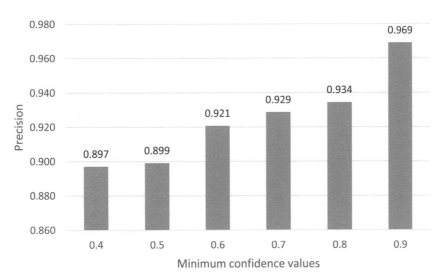

Figure 4.11   Precision of QARM with MovieLens dataset using fixed support = 0.3.

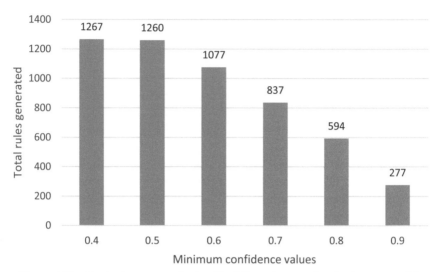

Figure 4.12   Total rules generated using MovieLens dataset using fixed support = 0.3.

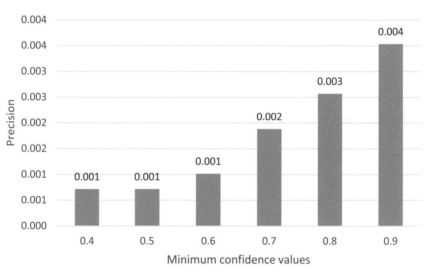

Figure 4.13   Precision of QARM with MovieLens dataset using fixed support = 0.35.

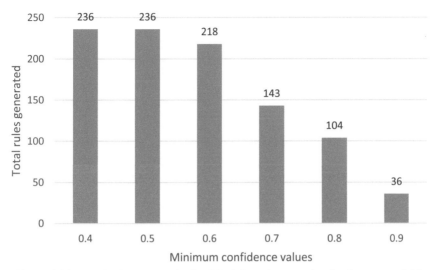

Figure 4.14 Total rules generated using MovieLens dataset using fixed support = 0.35.

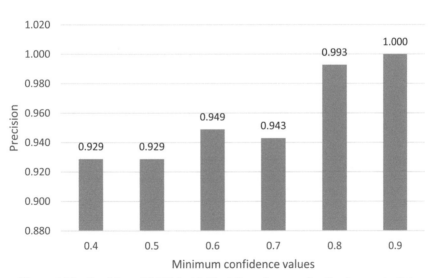

Figure 4.15 Precision of QARM with MovieLens dataset using fixed support = 0.4.

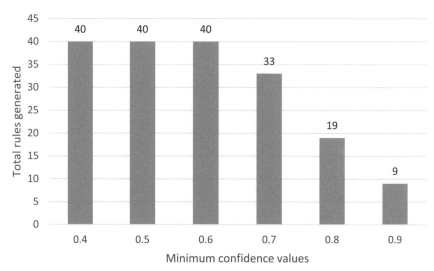

Figure 4.16   Total rules generated using MovieLens dataset using fixed support = 0.4.

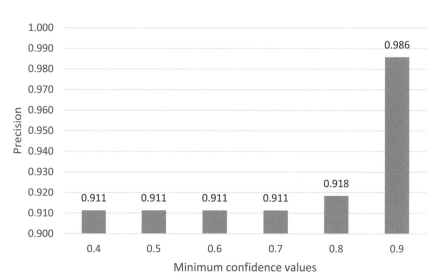

Figure 4.17   Precision of QARM with MovieLens dataset using fixed support = 0.45.

From the entire database containing the entire transaction history of all users, we filtered out all users (and their respective histories) that have size that is less than 15 transactions. This resulted in a total of 18,290 users, which we split into two subsets: (i) *test* set that equaled the 30% of the users and (ii) *training* set that equaled the 70% of the total users.

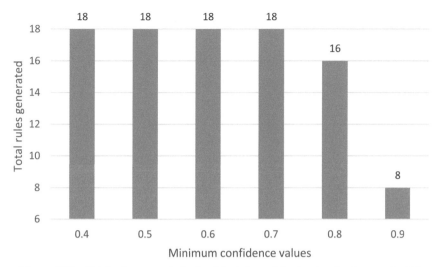

Figure 4.18 Total rules generated using MovieLens dataset using fixed support = 0.4.

Because of the nature of the dataset, and in order to generate some association rules, we had to lower the *support* value to as low as 0.5% and we generated a set of association rules under different confidence values (specifically, we experiment with confidence values {0.1, 0.2, 0.3, 0.4}). The number of association rules generated is depicted in Figure 4.19.

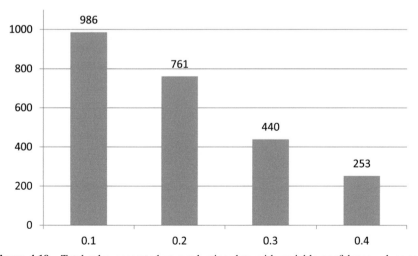

Figure 4.19 Total rules generated on production data with variable confidence values and fixed support.

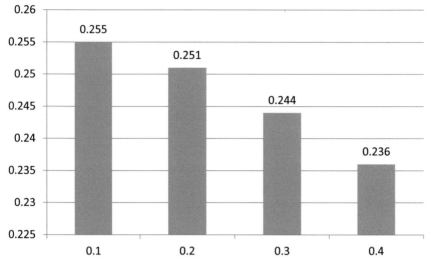

Figure 4.20   Recall value of the Post-Processor using generated association rules with fixed support and variable confidence values.

The introduced Post-Processor then used the generated association rules in order to perform the post-processing functionality on a set of generated recommendations by the hybrid recommender for all users in the *test* set. The different *recall* values of the enhanced recommender which combined both the hybrid recommender and the Post-Processor are depicted in Figure 4.20. As we showed in the experimental section of the previous chapter (on Recommender Systems) and specifically in Table 3.3 the recall value for the hybrid recommender (for $n = 10$) resulted in a recall value equal to *0.2*, which shows that the Post-Processing step improves the performance of the recommender in terms of recall.

5

Conclusions and Future Directions

In the current work, we presented different algorithms for academic search as well as recommender systems. The introduced algorithms share many common characteristics, the most important being that they aim to recommend content – be it *academic publications* or *consumable content* – that is considered to be *relevant* for users under specific parameters, including whether prior knowledge for a specific user exists. The introduced algorithms have high applicability in a range of settings and our implementation is fully modular so that transition to other settings may occur using the minimum possible required effort.

In the first part of the work, we have presented a novel ranking algorithm for academic search. We have presented a hierarchical heuristic scheme that aims to re-rank a set of results generated by third-party search engines in response to specific user-submitted queries. We have developed a meta-search engine that allows our heuristic scheme to generate alternative rankings of the original result set by taking into consideration different characteristics of the academic publications.

In order to measure the performance of the introduced scheme, we have performed a series of experiments against different commercial academic search engines including ACM Portal, Google Scholar, Microsoft Academic Search, and ArnetMiner and have used evaluations from a set of volunteers. Our experiments showed that for most queries, the introduced scheme outperformed many of the aforementioned search systems in terms of the score generated using different metrics.

The results of t-test, sign test, and signed rank test all indicate that PubSearch outperforms ACM Portal by a large and statistically significant margin in terms of all metrics considered, namely, lexicographic ordering LEX, NDCG, and ERR metrics. In terms of lexicographic ordering, *the average improvement* is in the order of 907.5%, in terms of NDCG, the *average improvement* is almost 12%, and in terms of the ERR metric, the *average improvement* is more than 77%. Similarly, comparing PubSearch against the standard Okapi BM25 scheme shows that PubSearch offers very significant advantages for ranking academic search results.

Even when comparing PubSearch against the current state-of-the-art academic search engines Google Scholar, Microsoft Academic Search, and ArnetMiner, the comparisons show that PubSearch outperforms these other engines in terms of all metrics considered, and in the vast majority of cases, by statistically significant margins.

Without detailed knowledge of the ranking system behind ACM Portal or the other academic search engines we compared our system with, we postulate that the main reasons for the better quality of our ranking scheme are in the *custom implementation of the term frequency heuristic* we have developed that takes into account the position of the various terms of a query in the document and the relative distance between the terms, as well as in the *chosen architecture itself*: the custom implementation of the term frequency score roughly determines whether a publication is relevant to a particular query, the time-depreciated citation distribution is a good indicator of the overall current value of a paper, and finally the clique score criterion accords extra value to (otherwise similarly ranked) papers that are classified in subjects that are strongly linked together as evidenced by the cliques formed in the Type I and II graphs that connect index terms together in a relatively small but typical publication base crawled for this purpose.

We have shown through extensive experimentation that the *proposed configuration outperforms all the other configurations we have experimented with, as well as the popular ensemble fusion approach using linear weights* (that is present in many if not most modern classifier, clustering, and recommender systems designed today, e.g., Christou et al., 2011).

In the second part of the work, we have presented a commercial, hybrid movie recommendation system that uses a novel ensemble of recommender algorithms of different types for improving performance in terms of recall. The implemented system addresses certain commonly encountered issues that many such commercial systems need to address, as well as many limitations originating from business specific requirements. The most significant issue that our system needed to address was to determine user preference given only information based on the items that have been consumed by members bound to a specific user account, in the absence of any relevance judgments.

Furthermore, the fact that a large percentage of all available content is currently offered at zero-price levels makes it easier for anyone to "purchase" items that they would not otherwise choose to purchase. And to add further complexity into this, without any information whatsoever concerning the percentage of the total playtime that the user actually watched any item, user histories can easily contain many consumed items which in fact not relevant at all to the user's true preferences.

By using different types of recommenders including *item* based, *user* based, as well as *content* based, the ensemble is able to handle cases where a significant number of users have consumed a significant number of items, thus taking advantage of the benefits of collaborative filtering, as well as cases where new content items have not been yet consumed by any user but by using its content-based recommender, the ensemble can still provide meaningful recommendations (thus, at least partially, solving the "new content" cold-start type problem).

The system also deals with the "new user" cold-start type problem (when new users are added to the system), using the following business rule: whenever a new user is inserted into the system, and before they have purchased any items, the system simply recommends the top recommendations made for all other users in the system at that time.

Finally, the system addresses hardware infrastructure constraints; we have introduced a cost-effective way with which the system is able to provide instant replies to web service requests and at the same time renew user recommendations as frequently as possible (in the order of 15 min or less). This was made possible by the architecture of the system, as well as the two databases following the same data model that are used to separate the updates of the system (performed by the AMORE batch job) from the response to web-service requests for recommendations (performed by the AMORE web-services that live in a web application server).

AMORE is currently the only live commercial recommender system for video-on-demand in Greece, and has been successfully deployed in the production environment of a prominent Greek Triple Play services' provider and has already contributed to an increase of the provider's profits in terms of movie rental sales, while at the same time offers customer retention support allowing the company's Marketing Department to offer more interesting subscription offers to both old and new customers alike.

We have experimented with the application of various algorithms implemented in the Apache Mahout suit-case (upon which myrrix is also based, see http://myrrix.com) but the results were not deemed satisfactory neither in terms of quality nor in terms of response times, thus necessitating the development of our own parallel multi-threaded custom implementation of the well-known k-NN-item-based and k-NN-user-based recommenders and variants thereof. Various other experimental recommendation systems have already shown the superiority of hybrid systems incorporating tens or even hundreds of individual recommender algorithms over schemes incorporating only a single algorithm (the best Netflix prize contestants belong in this category). AMORE results have shown that a very small number of different types of recommender algorithms (that can be updated very fast) are sufficient to produce high-quality recommendations that users enjoy: currently, the users make a rental from the proposed recommendations once for every two visits to the AMORE recommendations screen. In the immediate future, we are aiming to introduce novel algorithms which take into consideration additional information about user behavior patterns including the prices that users are willing to pay in order to provide improved recommendation services to them.

Finally, in the third part of the work, we have presented QARM, a novel algorithm for mining QARs. QARM mines such QARs by processing a large number of user histories in order to generate a set of association rules with a minimally required support and confidence value. The generated rules show strong relationships that exist between the consequent and the antecedent of each rule, representing different items that have been consumed at specific price levels. We are then using the aforementioned information as part of a post-processing mechanism that is used on top of the recommendation results generated by recommenders. Our experiments show that using the post-processor on top of the results generated by our introduced recommenders improves the original recommendation functionality.

Furthermore, since the introduced QARM algorithm has high applicability in a range of different e-commerce-related settings, we have performed a series of simulations under different datasets (with respect to size and complexity) and are characterized by different parameters such as the total number of users that exist, the available consumable items in the dataset, and the number of cycles during each every user consumed a pre-specified number of items, all of which represent different business scenarios. We then executed a series of experiments on the generated datasets to show the performance of the QARM algorithm.

References

Amolochitis, E., Christou, I. T., and Tan, Z. H. (2012). "PubSearch: a hierarchical heuristic scheme for ranking academic search results," in *Proceedings of the 1st International Conference on Pattern Recognition Applications and Methods*, Algarve.

Amolochitis, E., Christou, I. T., Tan, Z. H., and Prasad, R. (2013). A heuristic hierarchical scheme for academic search and retrieval. *Inform. Process. Manag.* 49, 1326–1343.

Attar, R., and Fraenkel, A. S. (1977). Local feedback in full-text retrieval systems. *J. ACM* 24, 397–417.

Beel, J., and Gipp, B. (2009). "Google scholar's ranking algorithm: an introductory overview," in *Proceedings of the 12th International Conference on Scientometrics and Infometrics*, Rio de Janeiro, 230–241.

Bron, C., and Kerbosch, J. (1973). Algorithm 457: finding all cliques of an undirected graph. *Communicat. ACM* 16, 575–577.

Cha, M., Kwak, H., Rodriguez, P., Ahn, Y. Y., and Moon, S. (2007). "I tube, you tube, everybody tubes: analyzing the world's largest user generated content video system," in *Proceedings of the 7th ACM SIGCOMM Conference on Internet Measurement,* (New York, NY: ACM), 1–14.

Chapelle, O., Metlzer, D., Zhang, Y., and Grinspan, P. (2009). "Expected reciprocal rank for graded relevance," in *Proceedings of the 18th ACM Conference on Information and Knowledge Management*, New York, NY, 621–630.

Cho, J., and Roy, S. (2004). "Impact of search engines on page popularity," in *Proceedings of the 13th International Conference on World Wide Web*, (New York, NY: ACM), 20–29.

Christou, I. T., Gkekas, G., and Kyrikou, A. (2011). A classifier ensemble approach to the TV viewer profile adaptation problem. *Intl. J. Machine Lear. Cybernet.* 3, 313–326. doi: 10.1007/s13042-011-0066-4

Czernicki, B. (2009). *Next-Generation Business Intelligence Software with Silverlight 3*. New York, NY: Apress.

Deshpande, M., and Karypis, G. (2004). Item-based top-n recommendation algorithms. *ACM Trans. Informat. Syst.* 22, 143–177.

Ekstrand, M. D., Ludwig, M., Konstan, J. A., and Riedl, J. T. (2011). "Rethinking the recommender research ecosystem: reproducibility, openness, and LensKit," in *Proceedings of the Fifth ACM Conference on Recommender systems*, (New York, NY: ACM), 133–140.

Garey, M. R., and Johnson, D. S. (1979). *Computers and Intractability: A Guide to the Theory of NP-Completeness*. San Francisco, CA: Freeman.

Golbeck, J., and Hendler, J. (2006). "Filmtrust: movie recommendations using trust in web-based social networks," in *Proceedings of the IEEE Consumer Communications and Networking Conference*, Vol. 96, (College Park, MD: University of Maryland).

Goldberg, D., Nichols, D., Oki, B. M., and Terry, D. (1992). Using collaborative filtering to weave an information tapestry. *Commun. ACM* 35, 61–70.

Harpale, A., Yang, Y., Gopal, S., He, D., and Yue, Z. (2010). "CiteData: a new multi-faceted dataset for evaluating personalized search performance," in *Proceedings of ACM Conference on Information and Knowledge Management CIKM,* Toronto, ON.

Herlocker, J. L., Konstan, J. A., Terveen, L. G., and Riedl, J. T. (2004). Evaluating collaborative filtering recommender systems. *ACM Trans. Informat. Syst.* 22, 5–53.

Hurley, N., and Zhang, M. (2011). Novelty and diversity in top-n recommendation – analysis and evaluation. *ACM Trans. Int. Technol.* 10:14.

Jackson, P., and Moulinier, I. (2002). *Natural Language Processing for Online Applications: Text Retrieval, Extraction and Categorization.* Amsterdam: John Benjamins.

Jahrer, M., Töscher, A., and Legenstein, R. (2010). "Combining predictions for accurate recommender systems," in *Proceedings of the 16th ACM SIGKDD International Conference on Knowledge Discovery and Data Mining*, (New York, NY: ACM), 693–702.

Järvelin, K., and Kekäläinen, J. (2000). "IR evaluation methods for retrieving highly relevant documents," in *Proceedings 23rd Annual International ACM SIGIR Conference on Research and Development in Information Retrieval*, (New York, NY: ACM), 131–150.

Karypis, G. (2001). "Evaluation of item-based top-n recommendation algorithms," in *Proceedings of the Tenth International Conference on*

Information and Knowledge Management, (New York, NY: ACM), 247–254.

Kirsch, S. M., Gnasa, M., and Cremers, A. B. (2006). "Beyond the web: retrieval in social information spaces," in *Proceedings of the 2006 European Conference on Advances in Information Retrieval* (Ann Arbor, MI: ECIR), 84–95.

Kotsiantis, S., and Kanellopoulos, D. (2006). Association rules mining: a recent overview. *GESTS Int. Trans. Comput. Sci. Eng.* 32, 71–82.

Krieger, U. D. (2011). An algebraic multigrid solution of large hierarchical Markovian models arising in web information retrieval. *Lect. Notes Comput. Sci.* 5233, 548–570.

Kuncheva, L. (2004). *Combining Pattern Classifiers: Methods and Algorithms.* Hoboken, NJ: Wiley.

Kuramochi, M., and Karypis, G. (2001). "Frequent subgraph discovery," in *Proceedings IEEE International Conference on Data Mining*, San Jose, CA, 313–320.

Lathia, N., Hailes, S., Capra, L., and Amatriain, X. (2010). "Temporal diversity in recommender systems," in *Proceedings of the 33rd International ACM SIGIR Conference on Research and Development in Information retrieval*, (New York, NY: ACM), 210–217.

Leung, C. W. K., Chan, S. C. F., and Chung, F. L. (2006). A collaborative filtering framework based on fuzzy association rules and multiple-level similarity. *Knowl. Informat. Syst.* 10, 357–381.

Li, Q., Chen, Y. P., Myaeng, S.-H, Jin, Y., and Kang, B.-Y. (2009). Concept unification of terms in different languages via web mining for information retrieval. *Informat. Process. Manag.* 45, 246–262.

Li, Y., Lu, L., and Xuefeng, L. (2005). A hybrid collaborative filtering method for multiple-interests and multiple-content recommendation in E-Commerce. *Exp. Syst. Applicat.* 28, 67–77.

Limbu, D. K., Connor, A., Pears, R., and MacDonell, S. (2006). "Contextual relevance feedback in web information retrieval," in *Proceedings of the 1st International Conference on Information Interaction in Context*, Copenhagen, 138–143.

Lin, W., Alvarez, S. A., and Ruiz, C. (2002). Efficient adaptive-support association rule mining for recommender systems. *Data Min. Knowl. Discov.* 6, 83–105.

Ma, N., Guan, J., and Zhao, Y. (2008). Bringing PageRank to the citation analysis. *Informat. Proc. Manag.* 44, 800–810.

Manning, C. D., Raghavan, P., and Schutze, H. (2009). *An Introduction to Information Retrieval*. Cambridge: Cambridge University Press.

Matsuo, Y., Mori, J., Hamasaki, M., Ishida, K., Nishimura, T., Takeda, H., et al. (2006). "Polyphonet: an advanced social extraction system from the web," in *Proceedings of the 15th International Conference on World Wide Web, WWW '06*, Edinburgh.

Martinez-Bazan, N., Muntes-Mulero, V., Gomez-Villamor, S., Nin, J., Sanchez-Martinez, M.-A., and Lariba-Pey J.-L. (2007). "DEX: high-performance exploration on large graphs for information retrieval," in *Proceedings ACM Conference on Information and Knowledge Management CIKM 07*, Lisboa.

Mild, A., and Natter, M. (2002). Collaborative filtering or regression models for Internet recommendation systems? *J. Target. Measur. Anal. Mark.* 10, 304–313.

Mobasher, B., Dai, H., Luo, T., and Nakagawa, M. (2001). "Effective personalization based on association rule discovery from web usage data," in *Proceedings of the 3rd International Workshop on Web Information and Data Management* (New York, NY: ACM), 9–15.

Owen, S., Anil, R., Dunning, T., and Friedman, E. (2011). *Mahout in Action*. Shelter Island, NY: Manning, 145–183.

Pazzani, M. J., and Billsus, D. (2007). "Content-based recommendation systems," in *The Adaptive Web*, (Berlin: Springer), 325–341.

Ricci, F., Rokach, L., and Shapira, B. (2011). *Introduction to Recommender Systems Handbook*. Berlin: Springer, 1–35.

Samudrala, R., and Moult, J. (1998). A graph-theoretic algorithm for comparative modeling of protein structure. *J. Mol. Biol.* 279, 287–302.

Sarwar, B., Karypis, G., Konstan, J., and Riedl, J. (2000). "Analysis of recommendation algorithms for e-commerce," in *Proceedings of the 2nd ACM Conference on Electronic Commerce* (New York, NY: ACM), 158–167.

Schaeffer, S. E. (2007). Graph clustering. *Comput. Sci. Rev.* 1, 27–64.

Shani, G., and Gunawardana, A. (2011). "Evaluating recommendation systems," in *Recommender Systems Handbook*, eds F. Ricci, L. Rokash, B. Shapira, and P. B. Kantor (Berlin: Springer), 257–297.

Shardanand, U., and Maes, P. (1995). "Social information filtering: algorithms for automating "word of mouth"," in *Proceedings of the SIGCHI Conference on Human Factors in Computing Systems,* (Boston, MA: Addison-Wesley Publishing Co.), 210–217.

Sarwar, B., Karypis, G., Konstan, J., and Riedl, J. (2000). *Application of Dimensionality Reduction in Recommender System – A Case Study*. Minneapolis, MN: University of Minnesota.

Sparck Jones, K., Walker, S., and Robertson, S. E. (2000). A probabilistic model of information retrieval: development and comparative experiments. *Informat. Process. Manag.* 36, 779–808.

Stoica, I., Morris, R., Liben-Nowell, D., Karger, D. R., Kaashoek, M. F., Dabek, F., et al. (2003). Chord: a scalable peer-to-peer lookup protocol for internet applications. *IEEE/ACM Trans. Netw.* 11, 17–32.

Tang, J., Jin, R., and Zhang, J. (2008). "A topic modeling approach and its integration into the random walk framework for academic search," in *Proceedings of 8th IEEE International Conference on Data Mining*, (Pisa: IEEE), 1055–1060.

Yu, H., Mine, T., and Amamiya, M. (2005). "An architecture for personal semantic web information retrieval system integrating web services and web contents," in *Proceedings of the IEEE International Conference on Web Services (ICWS 05)*, Orlando, FL, 329–336.

Yu, H., Zheng, D., Zhao, B. Y., and Zheng, W. (2006). "Understanding user behavior in large-scale video-on-demand systems," in *Proceedings of the 2006 EuroSys Conference: ACM SIGOPS Operating Systems Review*, Vol. 40 (New York, NY: ACM), 333–344.

Walters, W. H. (2007). Google scholar coverage of a multidisciplinary field. *Informat. Process. Manag.* 43, 1121–1132.

Wang, F. H., and Shao, H. M. (2004). Effective personalized recommendation based on time-framed navigation clustering and association mining. *Exp. Syst. Appl.* 27, 365–377.

Wong, C., Shiu, S., and Pal, S. (2001). "Mining fuzzy association rules for web access case adaptation," in *Proceedings of the Workshop Program at the Fourth International Conference on Case-Based Reasoning*. Washington, DC.

Index

About the Author

Emmanouil Amolochitis currently works as Lead Data Scientist at Cepal Hellas Financial Services SA, in charge of the Data Science and Analytics team, acting as principal technical expert, data architect and manager. In the past, Emmanouil has worked as: i) Head of R&D at VoiceWeb SA, ii) Principal Software Engineer / Architect at Ethnodata SA, iii) Senior Software Engineer at Intrasoft International SA, iv) Senior Research Associate at Athens Information Technology on commercial research projects for Hellas-On-Line (commercial movie recommendation engine) and Intralot SA/OPAP (fraud detection in games of chance and lotteries). Emmanouil holds a Ph.D. in Computer Science from Aalborg University, Denmark, and has authored and co-authored a number of papers in high-impact journals.